Urban Sustainability

Editor-in-Chief

Ali Cheshmehzangi , Architecture & Built Environment, University of Nottingham Ningbo China, Ningbo, Zhejiang, China

The Urban Sustainability Book Series is a valuable resource for sustainability and urban-related education and research. It offers an inter-disciplinary platform covering all four areas of practice, policy, education, research, and their nexus. The publications in this series are related to critical areas of sustainability, urban studies, planning, and urban geography.

This book series aims to put together cutting-edge research findings linked to the overarching field of urban sustainability. The scope and nature of the topic are broad and interdisciplinary and bring together various associated disciplines from sustainable development, environmental sciences, urbanism, etc. With many advanced research findings in the field, there is a need to put together various discussions and contributions on specific sustainability fields, covering a good range of topics on sustainable development, sustainable urbanism, and urban sustainability. Despite the broad range of issues, we note the importance of practical and policy-oriented directions, extending the literature and directions and pathways towards achieving urban sustainability.

The series will appeal to urbanists, geographers, planners, engineers, architects, governmental authorities, policymakers, researchers of all levels, and to all of those interested in a wide-ranging overview of urban sustainability and its associated fields. The series includes monographs and edited volumes, covering a range of topics under the urban sustainability topic, which can also be used for teaching materials.

Chao Ye · Liang Zhuang

Urbanization and Production of Space

A Multi-scalar Empirical Study Based on China's Cases

Chao Ye
East China Normal University
Shanghai, China

Liang Zhuang
East China Normal University
Shanghai, China

ISSN 2731-6483 ISSN 2731-6491 (electronic)
Urban Sustainability
ISBN 978-981-99-1805-8 ISBN 978-981-99-1806-5 (eBook)
https://doi.org/10.1007/978-981-99-1806-5

© The Editor(s) (if applicable) and The Author(s), under exclusive license to Springer Nature Singapore Pte Ltd. 2023

This work is subject to copyright. All rights are solely and exclusively licensed by the Publisher, whether the whole or part of the material is concerned, specifically the rights of translation, reprinting, reuse of illustrations, recitation, broadcasting, reproduction on microfilms or in any other physical way, and transmission or information storage and retrieval, electronic adaptation, computer software, or by similar or dissimilar methodology now known or hereafter developed.

The use of general descriptive names, registered names, trademarks, service marks, etc. in this publication does not imply, even in the absence of a specific statement, that such names are exempt from the relevant protective laws and regulations and therefore free for general use.

The publisher, the authors, and the editors are safe to assume that the advice and information in this book are believed to be true and accurate at the date of publication. Neither the publisher nor the authors or the editors give a warranty, expressed or implied, with respect to the material contained herein or for any errors or omissions that may have been made. The publisher remains neutral with regard to jurisdictional claims in published maps and institutional affiliations.

This Springer imprint is published by the registered company Springer Nature Singapore Pte Ltd.
The registered company address is: 152 Beach Road, #21-01/04 Gateway East, Singapore 189721, Singapore

Acknowledgements

The biggest feature of rapid urbanization especially in the countries like China is the production of space. We must reflect deeply on the current model and path of global urbanization, particularly in China. This book aims to use the production of space as a critical theory to rethink about the process of urbanization, pointing out the importance of multi-scalar empirical study. Although the works of Marx and Engels did not emphasize the spatial scale of geography, later Marxist geographers focused on spatial issues, including how to combine historical and geographical issues. In response to the problem of urbanization, the combination of multi-scalar humanities and technology using the theory of spatial production is actually both traditional and cutting-edge, full of theoretical exploration and highly realistic. Urbanization and spatial production are a multi-scalar process, with different scales corresponding to different research methods. We hope that people will value the questions or propositions we have put forward more than the explanations and answers we have given. In the face of the strong impact and challenges of globalization on urbanization, an in-depth understanding of the scale effects and complexity of space is not only at the core of urbanization, urban–rural planning, and territorial spatial planning in the new era, but also a major task of academic circles. Whether it is a theoretical or empirical study, the most important thing is to stick to and innovate our critical thinking.

The first ten chapters in this book are substantially reproduced from papers already published. Chapter 1 was first published as "Review on Studies on Production of Urban Space", in *Economic Geography (in Chinese)* volume 31, and we are indebted to the editor for permission to publish it here. Chapters 2 and 5 were published in the journal named *Sustainability* from MDPI, and the former belongs to the special issue on *Building Regional Sustainability in Urban Agglomeration: Theories, Methodologies, and Applications*; we gratefully acknowledge their open-source permission to reproduce these materials. Chapters 3, 7, and 8 that followed the Yangtze River Delta from multiple scales were all published in *Habitat International* (see volumes 66, 42 and 98, respectively); we are delighted with the permissions granted by the editorial department. Chapter 4 was written as a paper to appear in volume 29 of the *Journal of Geographical Sciences*; we sincerely thank the editorial board for permission to republish this paper. Chapter 6 was delivered at the paper session on *Spatiotemporal*

Analytics in Earth Science at the 2018 Annual Meeting of the American Association of Geographers, and it was subsequently published in *Land Use Policy*, under the editorship of Yansui Liu, who has kindly given permission to republish this material. Chapters 9 and 10 appeared in *International Development Planning Review* volume 41 and *Journal of Rural Studies* volume 62, and we would like to thank their general editors for permissions to reproduce these articles. Chapter 11 is contribution original to this book.

Last but not least, we thank Senior Editor Lydia Wang and Project Coordinator Saranya Devi Balasubramanian of Springer Nature for their efforts in planning and publishing this book. We would also like to thank a number of people who have contributed to the writing of this book. Professor Yanwei Chai of Peking University continuously inspired and supported our work in theoretical research. Professor Mingxing Chen of the Chinese Academy of Sciences is a helpful fellow of our long-term collaboration. Xiangyi Ma and Yang Gao proved to be diligent and capable collaborators. Finally, we would like to dedicate this book to all good committed friends everywhere.

Overview

This book studies China's urbanization based on the theory of production of space. According to reconsidering and redefining the concept of production of space, this book reconstructs a new theoretical framework for multi-scalar empirical research on urbanization. Urban agglomeration and regional development strategies, development zone policies, and rural construction movements are the main modes of promoting urbanization in China. Combining with the rapid urbanization in China since the reform and opening up, we choose the most representative real cases, such as urban agglomerations in the Yangtze River Delta, Jiangsu Province, the New Urban District, State-Level New Area, National High-tech Industrial Development Zone, a university town in Nanjing, a town and its villages in Shanghai, to conduct a series of multi-scalar empirical studies from the macro- to the meso- and then to the microscale. Through an in-depth analysis of the interaction of social space changes and urbanization influenced by power, capital, and class, the book reveals the essence of China's urbanization that is dominated by the logic of production of space. The authors of this book propose that China's urbanization in the future should achieve an important transformation toward humanism and emphasize the even and adequate development among regions and between urban and rural areas. This also provides new ideas for the theory and practice of urbanization in other countries around the world.

This book can be read and referenced by researchers in the fields of urban and rural studies, geography, sociology, urban and rural planning, and management. It can also be used as a teaching reference book for teachers, researchers, and students in related scientific research institutions.

Contents

1	**Urbanization and the Production of Space**		1
	1.1	The Theory on Production of Space	1
		1.1.1 Background	1
		1.1.2 The Production of Space	2
	1.2	The Production of Space in China	3
	1.3	A New Research Perspective on China's Urbanization	4
	1.4	Structure of the Book	5
	References		6
2	**Spatial Production and Governance of Urban Agglomeration in China**		9
	2.1	Introduction	10
	2.2	Methodology	11
	2.3	Spatial Production of the YRDUA	13
		2.3.1 Unbalanced Development of the YRDUA	13
		2.3.2 The YRDUA Driven by Capital and Power	15
		2.3.3 Growing Regional and Social Inequality: Urbanization, Capital, and Urban–Rural Disparities	20
	2.4	Conclusions and Discussions	26
	References		28
3	**Production of Space in the Middle-Scale Region in China**		31
	3.1	Introduction	32
	3.2	Methodology: The Evaluation Indexes on Production of Space	33
	3.3	The Evolving Characteristics of Production of Space in the Process of Jiangsu's Urbanization	35
		3.3.1 The Uneven Geographical and Developmental Base in Jiangsu	35
		3.3.2 The Evolving Trend of Jiangsu's Urbanization Driven by Capital and Power	37

		3.3.3	The Constantly Growing City-Scale Inequity: Urbanization, Investment and the Rural–Urban Gap	40
	3.4	Conclusions and Discussions		44
	References			46

4 Spatial Production and Spatial Dialectic of New Urban Districts in China ... 49

	4.1	New Urban Districts, Urbanization and Spatial Production	50
	4.2	Research Data and Methods	52
		4.2.1 Research Object and Description	52
		4.2.2 Research Framework and Method	54
	4.3	Characteristics of Spatial Production in China's NUD	56
		4.3.1 Representations of Space: Construction of Space by Power	56
		4.3.2 Spatial Practice: Shape of Space by Power	58
		4.3.3 Spaces of Representation: Reconstruction of Space by Power	61
	4.4	Logics of Spatial Dialectic in China's NUD	64
		4.4.1 "Trinity": Category and Connotation	65
		4.4.2 Spatial Production and Reproduction	66
	4.5	Conclusion	67
	References		68

5 Disorder or Reorder? The Spatial Production of State-Level New Areas in China ... 71

	5.1	SLNAs and Spatial Production	71
	5.2	Research Methodology	73
		5.2.1 The Unique Feature of SLNA	74
		5.2.2 Spatial Overlay: From Zone to Area and to District	76
		5.2.3 Different Types: Management Model and Spatial Pattern	76
		5.2.4 Bidirectional Process: Bottom-Up and Top-Down	77
	5.3	The Evolving Spatiotemporal Characteristics of SLNAs	78
		5.3.1 The Temporal Characteristics of SLNAs	78
		5.3.2 The Spatial Characteristics of SLNAs	80
	5.4	Discussions: The Tension Between Disorder and Reorder	83
	5.5	Conclusions	85
	References		87

6 Spatial Production of National High-Tech Industrial Development Zones in China ... 91

	6.1	NHTIDZs and Spatial Production	91
	6.2	A Framework Explaining the Interaction Between Government, NHTIDZs and Urbanization	93

	6.3	The Evolving Relations Between NHTIDZs and Urbanization	95
		6.3.1 The Temporal Characteristics of NHTIDZs	95
		6.3.2 The Spatial Characteristics of NHTIDZs	95
		6.3.3 The Relationship Between NHTIDZs and Urbanization	98
	6.4	The Government-Dominated Model of Production of Space	99
	6.5	Discussion	103
	6.6	Conclusion	104
	References		105
7	**Multi-scalar Separations: The Production of Space of University Towns in China**		109
	7.1	Introduction	109
	7.2	Theory and Framework	110
	7.3	Xianlin University Town: Its Land Use and Policies	112
		7.3.1 Context	112
		7.3.2 Xianlin University Town LUCC	114
		7.3.3 Policies and Their Influences on Xianlin Town's LUCC	116
	7.4	Multi-scalar Separations of Xianlin University Town	117
		7.4.1 Time-Scale Separations	117
		7.4.2 Space-Scale Separations	118
		7.4.3 Social Separations	119
	7.5	Production of Space Mechanism in Xianlin University Town	121
	7.6	Discussion and Conclusions	122
	References		123
8	**The Lost Countryside: Spatial Production of Villages in China**		125
	8.1	Introduction	125
	8.2	Study Area and Methodology	127
		8.2.1 Study Area	127
		8.2.2 Research Methodology	128
	8.3	The Lost Countryside	131
		8.3.1 Ideological Space	131
		8.3.2 Superficial Space: Culture Only on the Wall	134
		8.3.3 Everyday Life Space: Limited Organization and Unformed Culture	135
	8.4	Conclusions and Discussions	137
	References		139
9	**Marginalized Countryside in a Globalized City: Production of Rural Space of Wujing Township in Shanghai, China**		143
	9.1	Introduction	143
	9.2	Methodology	145
		9.2.1 Research Method	145

		9.2.2	A Framework on Urbanization and Production of Rural Space	145
	9.3	Production of Space in Wujing		147
		9.3.1	Wujing Town as a Case Study	147
		9.3.2	The Characteristics and Dynamics of Spatial Production in Wujing	149
		9.3.3	The Role of Power and Capital	150
		9.3.4	The Marginalized Villages	153
	9.4	Discussion		154
	9.5	Conclusions		156
	References			157
10	**The Countryside Under Multiple High-Tension Lines: A Perspective on the Rural Construction of Heping Village, Shanghai**			159
	10.1	Introduction		159
	10.2	Study Area and Research Methodology		161
		10.2.1	Study Area	161
		10.2.2	Research Methodology	164
	10.3	The Countryside Under Multiple High-Tension Lines		165
		10.3.1	Physical High-Tension Lines	165
		10.3.2	Power High-Tension Lines	166
		10.3.3	Capital High-Tension Lines	169
		10.3.4	Psychological High-Tension Lines	171
	10.4	Discussion		173
	10.5	Conclusion		175
	References			175
11	**The Essence of Production of Space and Future of Urbanization**			179
	11.1	Three Characteristics of the Theory of Production of Space		179
	11.2	Production of Space Is the Driving Force of Urbanization		180
	11.3	Multi-scalar Intertwined Production of Space		181
	11.4	The Future of Urbanization: From People-Oriented to Humanism		183

List of Figures

Fig. 2.1	The framework of spatial production and urbanization	12
Fig. 2.2	YRDUA and its location in China	14
Fig. 2.3	Built-up area, FAI and the evolution of urbanization rate of the YRDUA	18
Fig. 2.4	The variation of REI in the YRDUA and the evolution of FDI	18
Fig. 2.5	The evolvement of urban residents' income in the YRD	19
Fig. 2.6	The evolution of urbanization rate of the YRDUA	21
Fig. 2.7	The evolution of URIR of the YRDUA	22
Fig. 2.8	The evolution of FAI in the YRDUA	24
Fig. 2.9	The evolution of FDI in the YRDUA	25
Fig. 2.10	The evolution of the urban area of the YRD	26
Fig. 3.1	Jiangsu's location and its three parts	36
Fig. 3.2	Evolution of built-up area, fixed assets investment and urbanization rate of Jiangsu	37
Fig. 3.3	Evolution of real estate development and foreign direct investment of Jiangsu	38
Fig. 3.4	Evolution of urban income, rural income and their proportion of Jiangsu	38
Fig. 3.5	Evolution of urbanization rate of Jiangsu	41
Fig. 3.6	Evolution of fixed assets investment of Jiangsu	42
Fig. 3.7	Evolution of foreign direct investment of Jiangsu	43
Fig. 3.8	Evolution of built-up area of Jiangsu	44
Fig. 3.9	Evolution of urban–rural income proportion of Jiangsu	45
Fig. 4.1	Spatial distribution of 244 NUDs in China during 1993 to 2016	53
Fig. 4.2	The stage evolution of NUDs in China	54
Fig. 4.3	Spatial model of NUDs in China	55
Fig. 4.4	Kernel analysis of spatial layout of NUDs in China during 1993–2016	59

Fig. 4.5	The spatial distribution of distances between regulatory agency and upper administration center and the changing trend of spatial elements of NUDs in China during 1993–2016	61
Fig. 4.6	The spatial production of NUDs and its ternary dialectic	64
Fig. 5.1	The basic evolution of spatial production of SLNAs	75
Fig. 5.2	Annual increment about indicators of SLNAs	80
Fig. 5.3	The spatial distribution and indicators of SLNAs in China	82
Fig. 5.4	The evolution of average GDP of Provinces in the four Regions of China	84
Fig. 5.5	Comparison of GDP between the best two SLNAs	86
Fig. 6.1	The framework about interaction between government, NHTIDZ and urbanization	94
Fig. 6.2	The evolution of NHTIDZs and the growth rate of Urbanization Rate (UR) in China	96
Fig. 6.3	The annual increment of NHTIDZs built in the main provinces	96
Fig. 6.4	Kernel density analysis of NHTIDZs space layout in China	97
Fig. 6.5	The linear fit trend between the area of NHTIDZs and the built-up area of China (2004–2018)	98
Fig. 7.1	Framework on the relations among LUCC, policies and production of space	112
Fig. 7.2	Distribution of Xianlin university town	113
Fig. 7.3	LUCC of Xianlin university town from 2002 to 2012	114
Fig. 7.4	Curve of change degree of policies and total increment of land use from 2002 to 2012	115
Fig. 7.5	Some representations of time-scale and space-scale separations	118
Fig. 7.6	Distribution of illegal inns and cars in Xianlin university town	120
Fig. 7.7	Mechanism of production of space in Xianlin	121
Fig. 8.1	Location of Tangwan village in Shanghai	128
Fig. 8.2	Culture and spatial production	130
Fig. 8.3	Parking space in Tangwan village	132
Fig. 8.4	Unbalanced living environment in Tangwan village	132
Fig. 8.5	Peng's Garden house	133
Fig. 8.6	Some representations of superficial space	135
Fig. 8.7	Some representations of everyday life space	136
Fig. 9.1	The framework on the relation between production of space and urbanization	146
Fig. 9.2	Location of Wujing township in Shanghai	148
Fig. 9.3	Migrants and the local people of Wujing from 2010 to 2015	149
Fig. 9.4	Division of the road, river and forest land	150
Fig. 9.5	Frequency of use of the public amenities and types of residents' daily activities	151

List of Figures

Fig. 9.6	The trend of commercial housing price and per capita net income of rural households in Wujing	153
Fig. 10.1	Location of Heping village in Shanghai	162
Fig. 10.2	Heping Village's population changes from 2013 to 2016	162
Fig. 10.3	Houses under high-tension lines	165
Fig. 10.4	Comparison of river conditions between Heping Village (left) and Tangwan Village (right)	167
Fig. 10.5	2016 investment structure for village and town construction	169
Fig. 10.6	The investment structure of Heping Village	170
Fig. 10.7	Construction investment and population comparison of Wujing's villages	171
Fig. 10.8	The empty Senior Center (left) and Women's Center (right)	172

List of Tables

Table 2.1	Main national policies and points in the evolution of the YRDUA (2001–2017)	16
Table 2.2	Main provincial policies and points in the evolution of the YRDUA (2001–2016)	17
Table 3.1	Main indexes assessing production of space in the course of urbanization	35
Table 4.1	Types of spatial arrangement in NUDs	59
Table 4.2	Area scale of NUDs	61
Table 5.1	The major types of national developmental zone	75
Table 5.2	Basic information of SLNAs	79
Table 6.1	National documents on China's NHTIDZs	101
Table 7.1	Big events and developmental orientations on Xianlin from 2002 to 2012	115
Table 8.1	Changes of some administrative division in Tangwan since Qing dynasty	129
Table 9.1	Valid questionnaire information	145
Table 9.2	Comparison between Shanghai and Minhang District in 2015	147
Table 9.3	Minhang District rural residents have lower income and public facilities	154
Table 10.1	An integrated methodology on rural studies	164
Table 10.2	Participant information from the valid samples	164
Table 10.3	Criteria for the "beautiful countryside construction of Wujing" project	168
Table 11.1	Key points of methodology for multi-scale spatial production research	183

Chapter 1
Urbanization and the Production of Space

Abstract With the rapid advancement of urbanization since the 1990s, China's urban spatial form and structure have undergone new and significant changes. Accordingly, it also urgently needs new theories to explain and cope with. Beginning in the 1970s, Western social sciences and urban academia set off a wave of research on urban spatial production. In particular, the neo-Marxist urban school and the Marxist geography school have continued to discuss the issues of the production of urban space and have accumulated relatively rich theoretical achievements and practical experiences. Inspired by this, Chinese academic circles have also begun to pay attention to and widely introduce the theory of production of space in recent years and to use it to explain urban phenomena and problems in China. Because the theory is so complex and multi-disciplinary that it is necessary for us to introduce it into our studies.

1.1 The Theory on Production of Space

1.1.1 Background

The production of urban or rural space refers to the process of reshaping the city by political and economic elements and forces such as capital, power and class, thereby making urban or rural space its medium and product. Before World War II, the regional school represented by Hettner and Hartshorne inherited and carried forward the tradition of viewing geography as a "space" science since Kant, and the concept of space is a "filled container" [10, 15]. In the 1950s and 1960s, geographers of the positivist and spatial analysts tried to find spatial "patterns" or "laws" with the help of mathematical tools [9, 24]. They advocated the urban geographer Christaller's Central Place Theory, which is essentially a form of urban spatial morphology or

---

This chapter is based on [*Economic Geography*, Ye, C., Chai, Y., & Zhang, X. (2011). Review on studies on production of urban space. *Economic Geography*, 31(3), 409–413. (叶超, 柴彦威, 张小林. 2011. "空间的生产"理论、研究进展及其对中国城市研究的启示. 经济地理, 31(3), 409–413)].

geometry. The research paradigm of positivist geography and urban geography is actually a "fetishism" of space [23], because it attempts to avoid value judgments about space, thereby ignoring the individuals, political and social relationships that (re)shape "space". Therefore, positivist geographers cannot reasonably explain and respond to the series of social, political and economic crises with the urban and suburban spatial differentiation and segregation problems formed by the differentiation of race and social class that occurred in the capitalist world in the late 1960s. This has led to a growing gap from location theory and positivist geography to real-world spatial problems [22]. Furthermore, the Marxist theory, which is known for profoundly criticizing, analyzing, and revealing the essence of the capitalist political and economic system, has naturally entered the field of vision of geographers and urban researchers.

Since the 1970s, with the help of Marxist theory, the neo-Marxist school on urban studies has formed. Its main theory is production of space [17]. Although the theory of production of space does not focus solely on urban issues, urban (urbanization) issues are at the heart of it. As David Harvey argues, urbanization and the production of space are intertwined [14]. Many theorists also focus on the production of urban space and regard urban/city as their focus.

1.1.2 The Production of Space

The French Marxist thinker Henri Lefebvre is the founder of the theory of production of space. He is considered to represent the best work of Marxism in spatial research [22] and the origin of postmodern critical human geography [27]. His theory is the most solid, imaginative and detailed part of the theory of production of space [25]. Lefebvre [17] conducted a relatively comprehensive philosophical investigation on the concept of space and profoundly criticized the traditional view that space is only regarded as a container and a "field". On this basis, he focused his theory on the production of urban space and put forward the core idea of "(social) space is a (social) product", and a trinity theoretical framework was constructed to show this spatial production process: (1) "spatial practice" represents the social production and reproduction of cities and everyday life, (2) "representations of space" refers to conceptualized spaces, such as those governed by the knowledge and ideology of scientists, planners, and social engineers; and (3) "spaces of representation" means the space of "inhabitants" and "users," which is in a position of being dominated and negatively experienced [17, 18]. By Marxist analytical tools, starting from the reality of capitalist urban development, Lefebvre finally constructed the theory of production of space with the core view that urban space is the product and producing process of capitalist production and consumption activities.

The theory of production of space has aroused great attention and positive responses from Marxist geographers. Harvey quoted and disseminated Lefebvre's ideas earlier. He not only initially used Marxist theory to elucidate the relationship between social justice and cities but also developed the urbanization theory of capital.

Harvey [11–13] pointed out that the urban spatial organization and structure are the needs and products of capital production, and the suburbanization of the middle class and the decline of urban centers are the inevitable results of the contradictory effects of capital accumulation and class struggle. The use and development of the theoretical weapon of "the production of space" enabled Harvey to effectively make up for the shortcomings of Marxism in urban issues and spatial dimensions, thus greatly enriching the theoretical connotation of urban production of space. Soja [27] emphasized the influence of political power and ideology on the production of urban space and tried to get rid of the double shackles of "fetishism" and the abstraction of space to construct postmodern geography according to "critical regional studies". Castells [2, 3] was also deeply influenced by Lefebvre and proposed the concept of collective consumption, arguing that urbanization has made the individual consumption of urban laborers increasingly become socialized collective consumption mediated by the state, and the main reason for urban development and spatial evolution is the capitalist system. The struggle between labor and capital and between workers and capitalists makes urban space a space for the reproduction of labor, and those urban plans and policies that take care of the interests of capital are not necessarily in the interests of the broad masses of urban dwellers and the poor. Some other scholars such as Smith pointed out that the uneven development of the capitalist political economy is the center of spatial production [25, 26]. Gregory and Dear further extended and deduced Lefebvre's concept of the trinity of production of space, and Dear even viewed Lefebvre as a "potential postmodernist," which reflects the forward-looking and continuous nature of the theory of production of space [5, 7].

Introducing Marxist theory into geography and urban studies with comprehensive social, spatial, and political-economic analysis, critique, and theoretical construction, is the main theoretical contribution of the neo-Marxists such as Lefebvre, Harvey, Castells and Smith. This comprehensive and interdisciplinary research has not only expanded and enriched the Marxist theoretical system but has also advanced the theoretical development of geography and urban areas, with implications beyond the geographical field. As the famous Marxist geographer [22] pointed out, the interaction between Marxism and environmental and spatial knowledge provides a powerful theoretical explanation for the profound problems of human existence, which in turn provides the disciplinary force for its formation. Based on the development history of Western human geography in the past 40 years, it can be said that, although the research trend of "institutional and cultural turn" and postmodern geography appeared later, the theoretical upsurge of production of space has not retreated but has been integrated to many issues and areas [6, 21, 28, 29].

1.2 The Production of Space in China

Since the 1990s, China's urban spatial structure and form have undergone significant changes. With the development of globalization and information technology,

China has entered a stage of rapid urbanization. Irrespective of the spatial expansion, resource environment, and social problems caused by rapid urbanization [8] or the judgment that the dynamic mechanism of urbanization in the new era tends to be diversified [4, 20, 33], and that urban development has entered a transitional period [16], the underlying consensus is that in the past 20 years, urbanization and urban space reshaped by a variety of new elements and forces, have acquired new characteristics, structures and evolutionary dynamics.

Among the new elements or forces, the flow of transnational capital, the struggle of various powers, and the differentiation of social classes have an increasingly strong impact on urban space. For example, the Shanghai World Financial Center, the tallest building in the Chinese mainland, was built by more than 40 companies from Japan, the United States, and other countries with a total investment of more than 1 billion dollars, representing the domination of international capital over urban space [19]. The concentration of high-priced housing in the city center has caused middle- and low-income citizens to migrate to the suburban fringes, reflecting the social differentiation brought about by urban gentrification [30]. The large-scale renovation and demolition of the old cities not only caused the problems of historical and cultural heritage protection but also led to frequent conflicts between developers and the aboriginal people, such as the "nail house" incident that has been reported in many places. The super-large leisure and entertainment city similar to "Xintiandi" built in Shanghai, Hangzhou, Chongqing, and other places has completely changed the flesh-and-blood connection between people, streets, and shops. This kind of consumption culture is becoming a means to control the production of urban space [1].

From the above typical examples, we can see those political and economic elements such as capital, power and class are becoming more and more important for reshaping urban space. Cities have become not only the "centers" and "growth poles" of traditional industrial agglomeration, but also places and tools for capital, power, and class interests to compete. The urban space affected by the above is undergoing fundamental changes that are different from those in the past. The spatial isolation of urban areas from suburbs and rich and poor neighborhoods produced by high-end housing and the strong impact of new urban architecture on social activities show that urbanization is shifting from the "production in space" to "production of space" as Lefebvre said. These urban phenomena and problems urgently call for new theoretical explanations.

1.3 A New Research Perspective on China's Urbanization

After entering the twenty-first century, there has also been a boom on researches of production of space in Chinese academic circles. On the one hand, this shows that the theory researches already have a certain foundation and "atmosphere" in China. On the other hand, it also reflects the fundamental changes in Chinese urban space and the urgency to be explained and resolved. Due to the traditional advantages of geography in spatial issues, the Anglo-American geographers are the main force and elucidator to promote the evolution of the production theory of space. However, there is still a lack of systematic and in-depth theoretical thinking and prominent case studies on the production of urban space in China. As researchers of a socialist country under the guidance of Marxism, not only should we pay full attention to Marxist geography and the neo-Marxist urban school, but we should also actively explore the applicability of these theories in China [32]. We should also strengthen the study of the social and cultural geography of cities, looking at and studying urban space issues from multiple perspectives [31]. In addition, the development process of the theory of production of space heralds a trend of interdisciplinary research, so it is high time to have an active dialogue between geography and other disciplines, between Chinese scholars and Anglo-American geographers.

From the perspective both real life and theoretical development, the production of space in China has become a crucial academic issue that needs to be studied urgently. Chinese geography is facing a rare opportunity to combine Marxism, urban issues and geography closely and develop a new theory. Many emerging relevant empirical cases also provide methods and materials. On this basis, it has become an important academic task to study the process and driving mechanism of production of space in China. In a letter to Chinese geographers in 2016, Harvey also quotes Marx's saying "ruthless criticizing of everything existing" to his Chinese colleagues. Marxist geography should be criticized dialectically while expounding its important value and research content especially in Chinese complex discourse and practical contexts. This is the basis for reconstructing Marxist geography.

1.4 Structure of the Book

Production of space has a variety of ideas and theories, as well as many complex interpretations. At the same time, the relevant practical problems have both general manifestations and their own characteristics in different countries, regions and places. At the ideological and theoretical level, production of space comes from logical and abstract thinking and finally is endowed with a symbolic concept or picture. At the practical or realistic level, problems are often manifested in diversity, ambiguity and particularity. This is especially true for the theories like production of space. Understanding and reconstructing the theory of production of space is actually a process of traveling back and forth between different scales and cases. Only by going through this process do we have the possibility to recreate a theory.

New theories and practices are born from new problems in practice. Based on ever-changing social realities, discovering and asking new fundamental questions are equivalent to opening the door to new theories and new practices. Theory and practice are not two roads but two links on one road. In other words, there are two indispensable tracks in parallel. The point of the theory and practice of a discipline is not to clearly distinguish their boundaries or even to argue which is more important. Instead, what matters most is to find the point of intersection or divergence between theory and practice. Therefore, the key to understanding and reconstructing the theory of production of space is to discover bridges and even to erect and build bridges. Furthermore, the bridges are used to connect two territories that appear to be separate and divided but actually can and should be traversed. On the problem of the relationship between theory and practice, it is still the methodology that is crucial.

The "human-earth relationship" is the core proposition of geography. "Earth" is the environment where we exist, perceive and live, which is also the meaning of space. Therefore, "people and space" is the mapping of the "human-earth relationship." The theory of production of space focuses on space, evolving and developing in the midst of changes. It is shown in free thought, deep critique, and broad integration, rather than some inherent framework or pattern. Faced with the reality of China's ultra-high-speed, ultra-large-scale, and drastic changes in urbanization, many academic studies have focused on empirical analysis of the characteristics, processes and dynamics of urbanization. The research on the unique spatiotemporal characteristics of China's urbanization and its internal development logic is relatively insufficient, which is exactly where the production of space theory can have an advantage. The theory of production of space involves the intricate relationship between time and space, multiple elements and multiple subjects or objects. It pays attention not only to spatial phenomena but also to the excavation of the production of space logic behind it. This realistic and critical theory is a powerful weapon to deeply understand the practical behavior and operation law in the process of urbanization at the present stage and to thoroughly study urbanization.

The main purpose of this book is to reconstruct and localize the theory of production of space in conjunction with multi-scale cases in the process of China's urbanization. Taking this as a guide, we explain the theory of production of space at different levels in Chap. 1 of this book. Chaps. 2–10 take the scale as a clue and select a series of typical cases (including the Yangtze River Delta urban agglomeration, Jiangsu Province, New Urban District, State-level New Area, National High-Tech Industrial Development Zone, a university town, a Township and its villages in Shanghai) for empirical researches. The combination of "from theory to practice" and "from macro to micro scale" enables the content of this book to be gradually deepened and promoted in an orderly manner. In the end, according to excavating the essence of space under different appearances to abstract the production of space logic behind China's urbanization, we try to condense and reach the conclusions in Chap. 11.

References

1. Bao Y (2006) Consumer culture and the production of urban space. Acad Mon (5):11–13. [包亚明. 2006. 消费文化与城市空间的生产. 学术月刊, (5): 11–13]
2. Castells M (1977) The urban question (A. Sheridan, Trans.). The MIT Press, Cambridge
3. Castells M (1983) The city and the grassroots. Edward Arnold, London
4. Chai Y (2000) Urban space. Science Press, Beijing. [柴彦威. 2000. 城市空间. 北京: 科学出版社]

5. Dear M (2004) The postmodern urban condition (X. Li, et al. Trans.). Shanghai Educational Publishing Press, Shanghai. [迪尔. 2004. 后现代都市状况. 李小科, 等, 译. 上海: 上海教育出版社]
6. Gottdiener M (1985) The social production of urban space. University of Texas Press, Austin
7. Gregory D (1994) Geographical imaginations. Blackwell, Oxford
8. Gu C (2006) New trends of China's urban development. City Plann Rev (3):26–31. [顾朝林. 2006. 中国城市发展的新趋势. 城市规划, (3): 26–31]
9. Haggett P, Cliff A, Frey A (1977) Locational analysis in human geography. Wiley, New York
10. Hartshorne R (1958) The concept of geography as a science of space, from Kant and Humboldt to Hettner. Ann Assoc Am Geogr 48(2):97–108
11. Harvey D (1973) Social justice and the city. Edward Arnold, London
12. Harvey D (1982) The limits to capital. Blackwell, Oxford
13. Harvey D (1985) The urbanization of capital. Blackwell, Oxford
14. Harvey D (2006) Lefebvre and "The production of space" (X. Huang, Trans.). Foreign Theor Trends (1):53–56. [哈维. 2006. 列菲弗尔与《空间的生产》. 黄晓武, 译. 国外理论动态, (1): 53–56]
15. Hettner A (1983) Geography: its history, nature and methods (L. Wang, Trans.). The Commercial Press, Beijing. [赫特纳. 1983. 地理学——它的历史、性质和方法. 王兰生, 译. 北京: 商务印书馆]
16. Huang Z (2004) The production of urban space: a globalized Shanghai. Soc Res Q (Taiwan) 53:61–83. [黄宗仪. 2004. 都市空间的生产: 全球化的上海. 社会研究季刊(台湾), 53: 61–83]
17. Lefebvre H (1991) The production of space (N. Smith, Trans.). Blackwell, Oxford
18. Lefebvre H (2016) Space and politics (C. Li, Trans.). Shanghai People's Publishing House, Shanghai [列斐伏尔. 2016. 空间与政治. 李春, 译. 上海: 上海人民出版社]
19. Lin GCS (2007) Chinese urbanism in question: state, society, and the reproduction of urban spaces. Urban Geogr 28(1):7–29
20. Ning Y (1998) New process of urbanization: dynamics and features of urbanization in China since 1990. Acta Geogr Sinica 53(5):88–95. [宁越敏. 1998. 新城市化进程——90年代中国城市化动力机制和特点探讨. 地理学报, 53(5): 88–95]
21. Olds K (1995) Globalization and the production of new urban spaces: Pacific Rim megaprojects in the late 20th century. Environ Plan A 27(11):1713–1743
22. Peet R (2007) Modern geographical thought (S. Zhou, et al. Trans.). The Commercial Press, Beijing. [皮特. 2007. 现代地理学思想. 周尚意, 等, 译. 北京: 商务印书馆]
23. Quaini M (1982) Geography and Marxism. Blackwell, Oxford
24. Schaefer F (1953) Exceptionalism in geography: a methodological examination. Ann Assoc Am Geogr 43(3):226–249
25. Smith N (1984) Uneven development: nature, capital and the production of space. Blackwell, Oxford
26. Smith N, O' Keefe P (1980) Geography, Marx and the concept of nature. Antipode 12(2):30–39
27. Soja EW (2004) Postmodern geography: Reaffirming space in critical social theory (W. Wang, Trans.). The Commercial Press, Beijing. [苏贾. 2004. 后现代地理学: 重申批判社会理论中的空间. 王文斌, 译. 北京: 商务印书馆]
28. Thrift N, French S (2002) The automatic production of space. Trans Inst Br Geogr 27(3):309–335
29. Unwin T (2000) A waste of space? Towards a critique of the social production of space. Trans Inst Br Geogr 25(1):11–29
30. Wu F (2000) The global and local dimensions of place-making: remaking Shanghai as a world city. Urban Stud 37(8):1359–1377

31. Xu X, Yao H (2009) Reviews and new progresses in China's urban geographical studies since 1900. Econ Geogr 29(9):1412–1420. [许学强, 姚华松. 2009. 百年来中国城市地理学研究回顾及展望. 经济地理, 29(9): 1412–1420]
32. Ye C, Cai Y (2010) Formation and evolution of radical geography: a case study as James Blaut's "The Dissenting Tradition". Sci Geogr Sin 30(1):1–7. [叶超, 蔡运龙. 2010. 激进地理学的形成和演变——以《异端的传统》为例. 地理科学, 30(1): 1–7]
33. Zhang J, Wu F, Ma R (2008) Institutional transformation and reconstruction of China's urban space: establishing an institutional analysis framework for spatial evolution. City Plann Rev 32(6):55–60. [张京祥, 吴缚龙, 马润潮. 2008. 体制转型与中国城市空间重构——建立一种空间演化的制度分析框架. 城市规划, 32(6): 55–60]

Chapter 2
Spatial Production and Governance of Urban Agglomeration in China

Abstract Urban agglomeration plays an essential role in world urbanization. Urban agglomerations in developing countries like China, although have the same characteristics as the developed countries because of globalization, often show a more different and dynamic process. Urban agglomerations in China are generally dominated and planned as a mode of organizing and improving urbanization by the Chinese government; however, in different regions, urban agglomeration has different trajectories based on a different historical and geographical context. The paper applies a new social theory, production of space, into explaining the development and governance of urban agglomeration in China, which is effective and meaningful to help understand the developmental process of urban agglomeration and urbanization. The theory of spatial production focuses on the relation between society and space, in which "society" has a broad meaning and can be divided into three factors or parts: power, capital and class. This paper chooses YRD (the Yangtze River Delta) as a typical case and designs a simple index system to reflect the influences of these three factors on urban agglomeration. The governance of urban agglomeration will be indicated by national or regional policies analysis. According to such a synthesis method of index assessment, GIS (Geographic Information System), and policies analysis, we find Chinese urban agglomeration is a capital-intensive region, and the national policies tend to regard it as an intensive investment object. Planning and governance from top-down power have more influence than the market in the evolving process of urban agglomeration. There is a contradiction between spatial production and people-oriented urbanization, with the latter more important than the former, but China's urbanization often emphasizes the former. It is high time to link the techniques and methods such as GIS to the social theories like the production of space in urban studies.

This chapter is based on [*Sustainability*, Ye, C., Liu, Z., Cai, W., Chen, R., Liu, L., & Cai, Y. (2019). Spatial production and governance of urban agglomeration in China 2000–2015: Yangtze River Delta as a case. *Sustainability*, 11(5), 1343].

2.1 Introduction

Since the notion of urban agglomeration was put forward, it has been the focus of global scholars [2, 8, 9]. There are a wide variety of concepts regarding urban agglomeration, including megalopolis, metropolitan area, city-region, and mega-city regions, however, urban agglomeration hasn't reached a universal definition [1, 10, 11, 29, 38]. Generally speaking, urban agglomeration is described as "the city-region consists of the core city and its surrounding areas, such as its suburban areas and outskirts……is a part of the recentralization of state power" [39]. Although chengshiqun (urban agglomeration in Chinese) has also been defined poorly in Chinese, it has attracted more attention than ever, especially since 2000. There are some different perspectives on urban agglomeration in China, such as state rescaling, developmental policies, intercity relationships, regional governance, and administrative annexation [5, 24–27, 46]. Urban agglomeration is often thought of as a type of state reterritorialization and city-region governance [3, 4, 22, 23, 39, 44]. The scholars have been looking for a way to help understand the developmental process of urban agglomeration. The above studies on urban agglomeration, focus on more quantitative or policy analysis than theoretical explanation. In particular, there is a lack of critical and profound theory for urban agglomeration research. In addition, Chinese urban agglomeration is a mixed production, driven by the market, social mobility, and strong policies, so it cannot be explained by only one theory or the Western theories. It is necessary to transform some theories based on Anglo-American experiences into Chinese versions regarding urbanization and urban agglomeration.

Compared to the existing studies on urban agglomeration, production of space, as a social theory, is critical and down to earth, which provides us a necessary and significant perspective to rethink the changing urban agglomerations in China. However, many studies on the production of space are so abstract and complicated that we can hardly put them into practice. Therefore, we try to transform the theory of spatial production into a brief framework including three elements and assess it from some indexes responding to the three elements, which makes the theory easier to understand and operate. The production of space generally means the relation and interaction between capital, power, class, and space [41, 42]. Together with intensive public and private investment and special national policies in Chinese urban agglomeration, the large-scale transformation of land use and the astonishing extension of urban space continually emerge, which is the most remarkable characteristic of spatial production. The Chinese government has proposed urban agglomeration as one of spatial organizations or tools to improve urbanization. In the process of the spatial production of urban agglomeration, power and capital play a very important role, however, how they work is not clear. Therefore, it is necessary to understand and explain how urban agglomeration in China happens and develops, especially driven by the force of capital and power.

Out of most urban agglomerations or city-regions of China, the Pearl River Delta and the YRD have been the most studied and, often, to a more thorough degree than the others. The YRD region, as a relatively mature, typical, and complex urban agglomeration, accounts for a large proportion of these studies. The paper attempts to explain and discover the real process and dynamics of the YRDUA (Yangtze River Delta Urban Agglomeration) based on the theory of spatial production. Because urban agglomeration has been rapidly developed since 2000, we choose the middle-term temporal scale from 2000 to 2015 as the research period. The structure of this paper covers the research methodology of spatial production and urbanization, characteristics of spatial production of the YRDUA, and the conclusion.

2.2 Methodology

The theory of spatial production which integrates the theories on space into social theories, reveals the role of capital and power operation in the process of urbanization, and subsequently, develops into a new theory. According to this new theory, space is a tool and a product in the whole process of social (space) production [7, 42]. Lefebvre pointed out that the production of space is similar to the production of any kind of commodity [19]. Firstly, various forms of social processes intervene in the urban space and shape it, secondly, (reshaping space is also a material power, which can influence and restrict social life and actions in the city [20]. Space appears together with time and society, in which everything including social relations, happens in space and time, space is often regarded as a tool in the production and a product [14–16, 32]. Capitalism, as a kind of complicated social system, also produces its space or is produced by the mode of spatial production, which is a dialectic process [17, 18, 33].

Lefebvre thought that urban space had become the key to the maintenance of capital circulation and capitalist production relations because space was simultaneously producing social relations during the process of spatial production [21]. According to socio-spatial dialectic, the methodology of spatial production, urban space continuously (re)shapes social relations and processes at different temporal scales and social relations react to urban space [12, 31]. Based on this methodology, society can be divided into three parts, political, economic, and (narrow-sense) social, which correspond to three main factors of power, capital, and class, respectively [41, 42]. Power, mainly from government, drives policymaking and governance. Capital is the most important factor in economic activity, and it flows across temporal and spatial scales, then produces an uneven geographical landscape [13]. Class also plays a significant role. With the improving strengths of power and capital in the process of urbanization, there is an increasingly large gap among social groups. In general, urbanization, especially in China, is the process and result of interaction among these three factors and all forms of urban space, see Fig. 2.1.

Politics, economy, and society are three important components of spatial production, so it is quite significant to evaluate and quantify them scientifically. Previous studies mostly focused on the interpretation of the theory of spatial production and revealed some phenomena in a pure, critical theoretical way. However, there is always a lack of a reasonable index system to assess and understand spatial production. So,

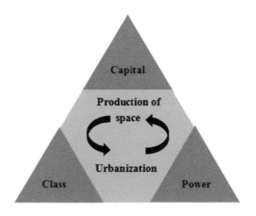

Fig. 2.1 The framework of spatial production and urbanization

from a statistical point of view, we can collect the corresponding statistical indexes to represent spatial production according to the above framework. Christophers [6] used two indicators with official data, fixed capital and pension fund based on the case of the United Kingdom. The first indicator can be used in our analysis, but as for the second, the more complicated indicator, considering the difference between China and the UK, we choose two indicators (fixed asset investment and foreign direct investment) to reflect capital, which are representative, because it will be more complicated if further indicators are adopted. Moreover, this article aims to demonstrate that capital and power work on urban agglomeration; therefore, besides capital, we should choose some other indicators to represent the role of power and society. There are many indicators and multi-data including mobile phone usage that can be chosen for measuring urban vitality and explored in research concerning the production of space [45]. However, in this paper, the new framework and integrating analysis, as opposed to data, are highlighted.

To illustrate the most important characteristics of spatial production and urbanization, this paper uses a simple index system, in which capital is represented by FAI (fixed asset investment) and FDI (foreign direct investment) and power is represented by the built-up area and the URIR (urban–rural income ratio), which can reflect class. The indicator of urbanization ratio reflects the level of urbanization [42]. In the developmental process of the YRDUA, the influences of capital globalization on the YRDUA and the evolving trends of internal spatial production can be demonstrated by indexes related to urban social and spatial extension. Here, we select built-up area, FAI, urbanization ratio, REI (real estate investment), FDI, and URIR as indexes to reflect the spatial and capital developmental process. In addition, the governance of urban agglomeration will be demonstrated by national policies and planning analysis. Therefore, the methodology of this article is blended and integrating. We not only design a brief and clear theoretical framework to make the theory easier to understand but also synthesize indexes, data, GIS, and policy analysis to represent the changing process of urban agglomeration. The data of this paper are mainly obtained from the China City Statistical Yearbook and Jiangsu, Anhui,

Shanghai, and Zhejiang Statistical Yearbook of 2016. All indicators except URIR are directly taken from the Yearbook, and URIR is calculated by the authors based on the data from the Yearbook. Although it is not sufficiently accurate, the official data are generally effective for large-scale regional analysis [42]. The paper is different from the general quantitative research on urbanization and its emphasis is not on statistical or model analysis. We try to design a new framework which links the theory of spatial production to the problems of urban agglomeration in order to reinterpret the developmental process and dynamics of urban agglomeration in China.

2.3 Spatial Production of the YRDUA

2.3.1 Unbalanced Development of the YRDUA

The YRDUA is the biggest comprehensive industrial base in China and has a prominent location that lies within the intersection of the Gold Coast of Eastern China and the Golden Waterway of the Yangtze River. Since 1992, the YRD has taken the lead in the process of regional integration reform, and global capital has started to be poured into the YRD. Since 2000, the YRD has become one of the regions with one of the highest economic growth rates and the fastest speed of urbanization in China, even in the world, which maintains the highest degree of urbanization and has the most densely distributed towns. Compared with urban agglomerations in other countries, the speed and scale of urbanization in the YRD also rank high on the list [40]. The population of the YRDUA exceeded 200 million in 2016, which was close to the world-class urban agglomerations of North America, Western Europe, and Japan. The spatial scale of the YRDUA has undergone a historical process of development which changes from a geographical concept to an economic concept and extends from small scale to large scale. According to the Development Planning of the YRDUA issued by National Development and Reform Commission, the YRDUA covers Shanghai, and Nanjing, Wuxi, Changzhou, Suzhou, Nantong, Yancheng, Yangzhou, Zhenjiang, and Taizhou of Jiangsu Province, Hangzhou, Ningbo, Jiaxing, Huzhou, Shaoxing, Jinhua, Zhoushan, and Taizhou of Zhejiang Province; and Hefei, Wuhu, Ma'anshan, Tongling, Anqing, Chuzhou, Chizhou, and Xuancheng of Anhui Province, see Fig. 2.2.

During the process of reform and opening up in the past 40 years, the economy of the YRD has been greatly developed. The GDP of the YRD reached 12.9 trillion yuan in 2014, accounting for 20.25% of the total country's GDP and was far higher than the national average level and other domestic urban agglomerations. Shanghai is dominant and has become one of the most important cities in the world. However, with the continuous spread of globalization, the development of cities in the YRD not only shows the trend of regional integration and spatial pattern of a multi-core network but also forms some metropolitan areas with the core of megalopolis like Shanghai, Nanjing, and Hangzhou. In 2015, the GDP per capita of Shanghai was

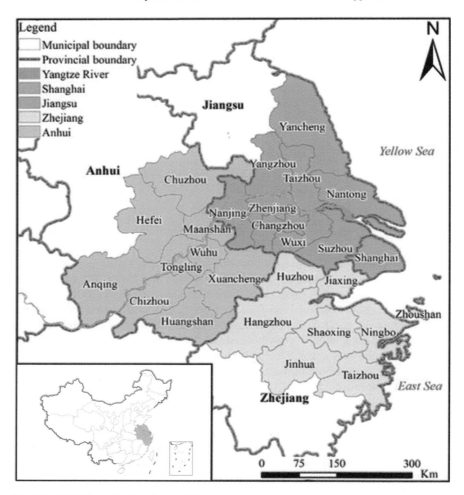

Fig. 2.2 YRDUA and its location in China

125,002 yuan; Zhejiang was 77,644 yuan; Jiangsu was 87,995 yuan, and Anhui was 35,997 yuan. The GDP per capita of Shanghai was over three times the GDP per capita of Anhui province and was about 1.5-times those of the Zhejiang and Jiangsu Provinces. The uneven distribution of resources, environment, capital, and policies leads to the imbalance of urban economic development inside the YRDUA, which has become a problem that cannot be ignored.

Uneven development is the core issue in understanding the production of space. Uneven development is the systematic geographical expression of the contradictions of the capitalist system [30]. Here, geographical conditions are both base and dynamic of regional or national development, which have been changed and reshaped in the production of space [42].

2.3.2 The YRDUA Driven by Capital and Power

During the evolution process of the YRDUA, Chinese national departments have issued relevant documents to guide them in recent years. From this perspective, building the YRDUA into a world-class urban agglomeration is not only the development orientation of itself but also Chinese governments' aim, which means power plays a crucial role in shaping and promoting the YRDUA.

During the years 2001–2016, sets of documents regarding the development of urban agglomeration were issued by different departments, see Tables 2.1 and 2.2, and the idea of taking urban agglomeration as the main form of future urban development was put forward by almost all of the documents. According to the documents for the development of the YRDUA, the common orientation is to form it into an important economic growth pole and then a world-class urban agglomeration with international competitiveness by promoting and optimizing international cooperation and competition at a higher level.

There are some remarkable differences among these documents regarding the orientation of the YRDUA. From the main national policies, "the 11th 5-Year plan" in 2006 first put forward the concept of urban agglomeration and regarded it as the main form of urban development to give priority to regional and national economic issues. "The 13th 5-Year plan" was based on this concept but came up with a deeper idea of internationalization, to gradually build some world-class urban agglomerations with more international and comprehensive competitiveness. After "the 12th 5-Year plan" had achieved some aims, the pursuit of faster and greater development was put forward by "the 13th 5-Year plan" which included optimizing and enhancing the function of the YRDUA. Under the guidance of Communist Party of China Central Committee's relevant documents, the departments including the State Council, the National Development and Reform Commission, and the Ministry of Housing and Urban–Rural Development all issued relevant documents which were more detailed and provided targeted supplementary and improvement in the view of the new situation and different urban agglomerations. Overall, although the contents of documents in each period are different, all departments expressed great expectations with regard to the development of the YRDUA, while it aimed at constructing a world-class urban agglomeration and playing a leading role in the national economic and social development through accelerating the formation of new advantages in international competition [43].

Over 15 years, FAI and FDI of the YRDUA have a similar changing trend, and the urbanization ratio basically shows a trend of linear growth. The evolution of built-up area and REI have a strong correlation (see Figs. 2.3 and 2.4). According to the relevant indexes, the development of the YRDUA can be divided into three periods, in which the year 2005 (the 11th Five-Year plan) and 2010 (the 12th Five-Year plan) are the turning points. The period of the years 2000–2005 can be referred to as the embryonic stage. There is an overall increasing trend but the development trend of each index remains relatively steady. This period can be described as the rapid development stage, in which each statistical index increases rapidly. The period

Table 2.1 Main national policies and points in the evolution of the YRDUA (2001–2017)

Year	Documents	Department	Orientation and key points
2001	The 10th 5-year plan	CPCCC	• Economic region • Economy-facilitated effects
2006	The 11th 5-year plan	CPCCC	• Urban agglomeration • Exert the leading and radiation actions • Reinforce the overall competitive power
2010	The 12th 5-year plan	CPCCC	• Promote economic integration • Perfect the layout • Plan scientifically • International competitiveness
2015	The 13th 5-year plan	CPCCC	• World-class urban agglomeration • Global competitiveness
2008	Further promoting reform and development in YRD	State Council	• International gateway to the Asia–Pacific Region • Global advanced manufacturing base • World-class urban agglomeration • International competitiveness
2014	Promoting YREB's development base on golden waterway	State Council	• Promote integrated development • World-class urban agglomeration • International competitiveness
2017	The outline of the national land plan (2016–2030)	State Council	• World-class urban agglomeration • International influence

Note Communist Party of China Central Committee (CPCCC); the Yangtze River Delta (YRD); Yangtze River Economic Belt (YREB). Date Source: 2017.6.20 search by: http://www.npc.gov.cn/wxzl/gongbao/2001-03/19/content_5134505.htm; http://www.npc.gov.cn/wxzl/gongbao/2006-03/18/content_5347869.htm; http://www.npc.gov.cn/wxzl/gongbao/2011-08/16/content_1665636.htm; http://www.npc.gov.cn/wxzl/gongbao/2016-07/08/content_1993756.htm; http://www.gov.cn/zhengce/content/2008-09/16/content_1715.htm; http://www.gov.cn/zhengce/content/2014-09/25/content_9092.htm; http://www.gov.cn/zhengce/content/2017-02/04/content_5165309.htm

of 2010–2015 is the sharp expansion stage when each index overall increases at an exponential speed, which is different from the previous two stages. For example, FAI and FDI have shown a strong momentum of growth. Although the overall evolution of each statistical index shows such a trend, there are some differences during certain separate years. For example, the scale of FAI did not increase but declined in 2005. Similarly, REI of the YRDUA in 2005 was basically the same as that in 2004 with no increase, as shown in Figs. 2.3 and 2.4. The underlying causes of the index changes will now be analyzed and explained.

Domestic and foreign capital investment which includes FAI, REI, and FDI, basically show similar variations to the urbanization ratio. After China joined the WTO (World Trade Organization) in 2001, FDI played a more important role than ever in

2.3 Spatial Production of the YRDUA

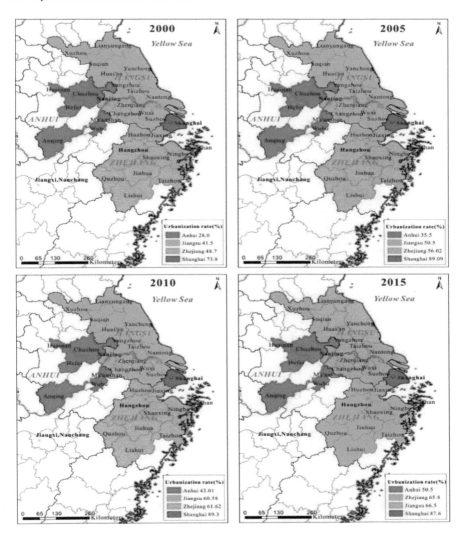

Fig. 2.6 The evolution of urbanization rate of the YRDUA. *Data source* [28, 34–37]

YRDUA. For the YRD as a whole, the average URIR is becoming smaller which means that the income gap between urban and the rural areas has widened. However, there has been an evident difference between all provinces in the YRD, see Fig. 2.7. Because of the high urban–rural coordination based on economic development, the difference of URIR between Shanghai, Zhejiang, and Jiangsu were smaller compared to that of capital and power, and the index of URIR was relatively low. However, the URIR of Anhui has been at a relatively high level constantly, as the highest among the provinces, though it narrowed from 2.74 in 2000 to 2.49 in 2015. Due to the level of natural conditions, economic and social development, and other factors, China's urbanization showed obvious regional differences. The gap was gradually widening, and indicated the uneven development inside each urban agglomeration and provincial regions.

The index of FAI is used to show the influence of capital on the urbanization process of the YRDUA. From 2000 to 2015, the amount of FAI in almost every city has been growing gradually, see Fig. 2.8. However, there were evident differences between different parts of the YRD. Cities around the Yangtze River basin

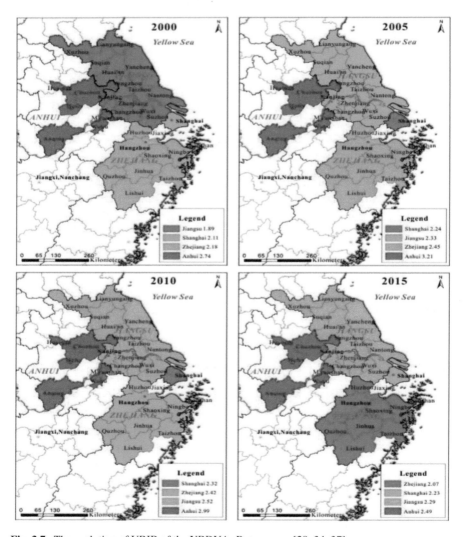

Fig. 2.7 The evolution of URIR of the YRDUA. *Data source* [28, 34–37]

2.3 Spatial Production of the YRDUA

have the highest FAI followed by Jiangsu, Zhejiang, and Anhui province. During this stage, Shanghai was the leading city in the amount of FAI while Tongling in Zhejiang province was essentially at the bottom. Moreover, the gap of FAI between cities rose by almost 20 times, from 394,678 billion Yuan in 2000 (Shanghai as the maximum 185,671, Zhoushan as the minimum 1807) to 8081.367 billion Yuan in 2015 (Shanghai as the maximum 60.297, Tongling as the minimum 66.508). The FAI of 13 cities in Jiangsu province was relatively high, which ranged from 106.927 billion Yuan in 2000 to 4155.275 billion Yuan in 2015, while the FAI of six cities in Anhui province ranged from 20,569 billion Yuan to 1276.82 billion Yuan. This fact indicates that cities in different provinces or in the same province have huge differences in their capital investment which lead to their urbanization ratio difference, see Fig. 2.6.

The FDI can directly indicate the influences of foreign capital in the process of urbanization in the YRDUA. According to the spatial evolution of the FDI in the YRDUA from 2000 to 2015, see Fig. 2.9, there is a large and obvious gap in FDI between cities around the Yangtze River basin and other cities in the Anhui, Zhejiang, and Jiangsu provinces; however, the FDI of all cities in the entire YRD region constantly increased in this stage. Shanghai and Suzhou have played a leading role since 2000. Taking cities around the Yangtze River Basin as the center, the cities of Zhejiang and Jiangsu province, close to Shanghai, have gradually expanded the scale of FDI since 2005, mainly because of a superior geographical location and convenient traffic. In 2000, the FDI of Shanghai was 6390 billion dollars, soaring to 18.166 billion dollars in 2015. In 2000, the city with the lowest FDI of only 8.69 million dollars was Zhoushan in Zhejiang province, and its FDI reached 200 million dollars in 2015. The difference in FDI between Shanghai and Zhoushan narrowed from 735 to 90 times. In 2015, Nanjing and Wuxi also ranked among those with the largest amount of FDI after Shanghai and Suzhou.

The spatial distribution and evolvement of FDI have generally aligned with the changing trends of urbanization and have basically remained unchanged since 2000. Since China joined WTO in 2001, developed countries such as the USA have also increased their investment in the YRD. More and more industries and factories have been investing, and the foreign investment has been infiltrating with increasing speed into the service industry such as commerce, securities, insurance banks, and other industries. The uneven distribution and growth of FDI in the inner urban agglomeration is one of the main causes of the increasing gap in regional economic development. From the distribution map of FDI in 2015, when capital cannot accumulate and grow in one region, it will find another region to absorb and digest surplus capital using a variety of mechanisms. This will further expand the uneven development in the urban agglomeration.

The index of built-up areas demonstrates the influence of power in the YRDUA. The built-up area has been rapidly expanding at a large scale in almost every city of the urban agglomeration every year. The regional and spatial differences are mainly shown in two aspects, see Fig. 2.10. Firstly, the built-up area of four cities including Shanghai, Hefei, Nanjing, and Hangzhou has expanded sharply. Secondly, the differences between cities in different provinces or in the same province increased

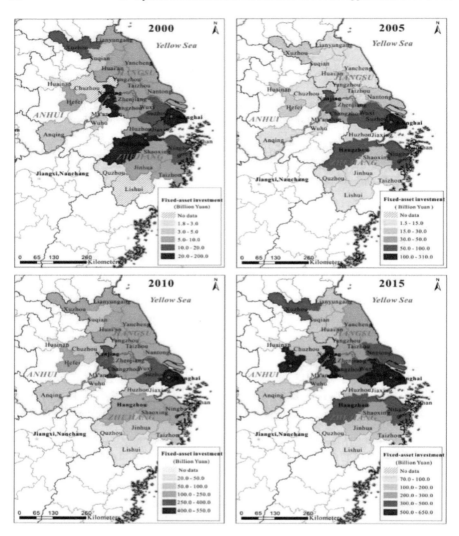

Fig. 2.8 The evolution of FAI in the YRDUA. *Data source* [28, 34–37]

during the years 2000–2015. The difference range increased from 3690 km^2 in 2000 (Shanghai as the maximum 550 km^2, which is almost 32 times that of the minimum, Tongling at 17 km^2) to 6452 km^2 in 2015 (Shanghai with the maximum 935 km^2, which is over 27.5 times that of Tongling with the minimum of 34 km^2).

In some cities, such as Shanghai, Hefei, Nanjing, and Hangzhou, the changing trends in their built-up areas are similar to that of the YRD overall. Shanghai has the built-up area which has expanded the most; the built-up area increased by 385 km^2, from 550 km^2 in 2000 to 935 km^2 in 2015. The built-up areas of Hefei, Nanjing, and

2.3 Spatial Production of the YRDUA

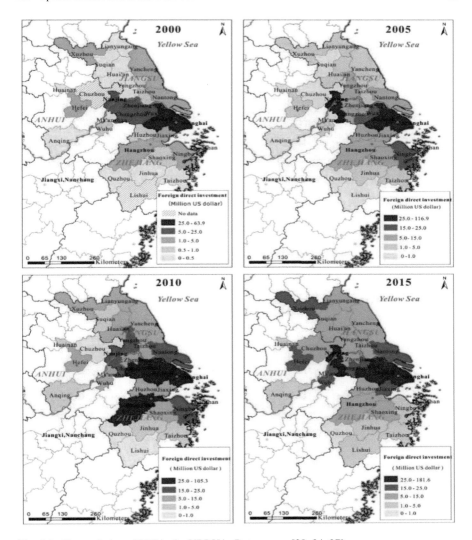

Fig. 2.9 The evolution of FDI in the YRDUA. *Data source* [28, 34–37]

Hangzhou have increased by 283, 540, and 324 km^2, respectively. The development of the YRDUA and the acceleration of urbanization are in the form of scale expansion. At first, it is influenced by capital investment including domestic investment and foreign investment. The curve of foreign investment has been continuously rising, although the growth rate of REI has been less than that of FDI and has fluctuated in some years. Moreover, it is affected by spatial expansion, which is reflected in the increasing built-up area. As the central cities of the YRD, Shanghai, Hefei, Nanjing,

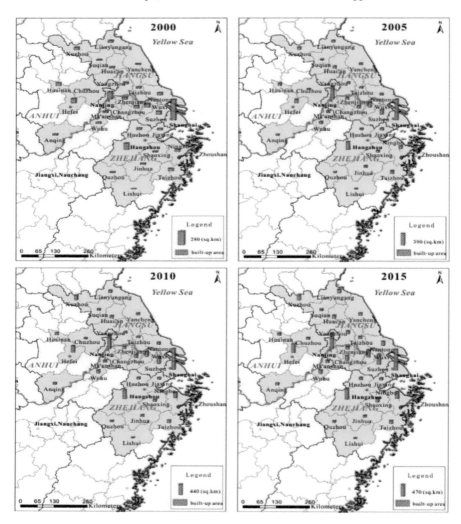

Fig. 2.10 The evolution of the urban area of the YRD. *Data source* [28, 34–37]

and Hangzhou are more powerful than other cities, which helps ensure the priority of expanding their built-up areas.

2.4 Conclusions and Discussions

Space, society, and time are interactively based on socio-spatial dialectic which focuses on the interactions and relations between time, space, and society at different

2.4 Conclusions and Discussions

scales [32, 41, 42]. In general, spatial production is a critical theory that should be applied to multi-scalar urbanization issues including urban agglomeration. The YRDUA has the free marketing history and base as shown by Jiangnan culture in Chinese history, but nowadays it needs to be intervened and designed by the national planning policies in the new period. The YRDUA has become a mixture dominated by governments at different levels together with its own geographical conditions and marketing force and as a result of spatial production driven by capital and power.

The evolution of indexes including domestic and foreign investment, built-up area, and the urbanization ratio of cities in the YRD generally have shown an increasingly upward trend, although some indexes have fluctuated. Observed from the indexes of capital and land use, the development of the YRD is roughly in the form of two axes. The main axis is the line from Shanghai, Suzhou-Wuxi-Changzhou, and Nanjing to Hefei, and the second axis is the link between Shanghai and Hangzhou. It shows a trend that is more toward centralization. Therefore, the development of the YRDUA strengthens the areas with advantages while the weak regions are still relatively weak and even weaker than before. From this view, the YRDUA has not yet played the role of a world-class urban agglomeration. The radiation of a central city is not enough, and the ties between areas are deficient. Furthermore, the functional division of cities in the YRDUA is often limited within the boundary of provinces.

While analyzing from spatial production, power (represented by built-up area and national policies), and capital (represented by domestic and foreign investment) has a great effect on the development of the YRDUA. The developmental orientation has undergone two transformations, and the years 2005 and 2010 are the turning points. Some important national plans from "the 10th Five-Year plan" to "the 13th Five-Year plan" have expressed the government's will toward the development of the YRD. It will be hard to change in the future. Although we analyze one urban agglomeration, it is probably similar to other regions or urban agglomerations in China.

This kind of traditional urbanization pattern, mainly driven by governmental power and characterized by large-scale spatial and capital extension, is almost contradicted with the aim of people-orientation. In "the 13th Five-Year plan", people-oriented urbanization as the core idea of a new type of urbanization is emphasized many times. In fact, individual or community development is really the core of people-oriented urbanization; however, the pattern of spatial production often overlooks and destroys it. Regional and social uneven development is reshaping and expanding with spatial production, and it seems to be hard to avoid the process and result. Urban agglomeration is always regarded as a tool or means for improving China's urbanization, which means spatial need is given priority over the right to city. Consequently, individual or local development will be inevitably ignored or even forgotten. In fact, the plans and policies on the large-scale region such as urban agglomeration focus on more spatial and capital extension rather than place-based development. On one hand, the strong logic of capital and power provides the driving force for rapid urbanization and helps achieve great economic development; on the other hand, spatial production dominated by this logic inevitably has a passive effect on people-oriented and place-based urbanization because the nature of the logic is pursuing more space and profit. If China's new-type urbanization in the future cannot change the pattern and the logic fundamentally and place more emphasis on individual or local development in the process of urbanization, it is impossible to achieve the aim of people-oriented urbanization.

Urbanization or urban agglomeration is a cross-disciplinary issue, which needs to introduce and integrate the different new theories, methodologies, and methods. We have indicated what the new social theory, production of space, is and how it can be applied in urban agglomeration research. The brief framework we design could help approach and transform the difficult theory into research, and the integration of qualitative and quantitative analysis could be helpful to represent the developmental process and dynamics of urban agglomeration. According to the profound theory and clear framework of spatial production, urban agglomerations, especially in China, are reinterpreted. Because of the complexity of the theory and urban issues, we simplify them and carry out some indicators and policy analysis, aiming to call for more attention to be given to the interactions and tensions between space and society in the process of rapid urbanization in China, which are becoming more and more important and serious. Some problems, such as the detailed role of capital and power, and accuracy of indicators, are still unsolvable, which are worth studying further. Anyway, rethinking society and reordering urbanization will be two main missions in the future, not only for the researchers but also for practitioners.

References

1. Baigent E (2004) Patrick Geddes, Lewis Mumford and Jean Gottmann: divisions over 'Megalopolis.' Prog Hum Geogr 28(6):687–700
2. Batten DF (1995) Network cities: creative urban agglomerations for the 21st century. Urban Stud 32(2):313–327
3. Brenner N (1999) Globalisation as reterritorialisation: the re-scaling of urban governance in the European Union. Urban Stud 36(3):431–451
4. Brenner N (2004) Urban governance and the production of new state spaces in Western Europe, 1960–2000. Rev Int Polit Econ 11(3):447–488
5. Chen M, Lu D, Zha L (2010) The comprehensive evaluation of China's urbanization and effects on resources and environment. J Geog Sci 20(1):17–30
6. Christophers B (2011) Revisiting the urbanization of capital. Ann Assoc Am Geogr 101(6):1347–1364
7. Cresswell T (2013) Geographic thought: a critical introduction. Wiley-Blackwell, Chichester
8. Duranton G, Puga D (2004) Micro-foundations of urban agglomeration economies. In: Henderson JV, Thisse JF (eds) Handbook of regional and urban economics, vol 4. Elsevier, Amsterdam, pp 2063–2117
9. Geddes P (1915) Cities in evolution: an introduction to the town planning movement and to the study of civics. Williams & Norgate, London
10. Gottmann J (1957) Megalopolis or the urbanization of the Northeastern Seaboard. Econ Geogr 33(3):189–200
11. Hall P, Pain K (2006) The polycentric metropolis: learning from mega-city regions in Europe. Earthscan, London
12. Harvey D (1973) Social justice and the city. Edward Arnold, London
13. Harvey D (1982) The limits to capital. Blackwell, Oxford
14. Harvey D (1985) The urbanization of capital. Blackwell, Oxford

References

15. Harvey D (1996) Justice, nature and the geography of difference. Blackwell, Cambridge
16. Harvey D (2000) Space of hope. Edinburgh University Press, Edinburgh
17. Harvey D (2003) Paris, capital of modernity. Routledge, New York
18. Harvey D (2012) Rebel cities: from the right to the city to the urban revolution. Verso, London
19. Lefebvre H (1991) The production of space (Smith N, Trans.). Blackwell, Oxford
20. Lefebvre H (1996) Writings on cities. Blackwell, Oxford
21. Lefebvre H (2003) The urban revolution. University of Minnesota Press, Minneapolis
22. Li Y, Wu F (2012) The transformation of regional governance in China: the rescaling of statehood. Prog Plan 78(2):55–99
23. Li Y, Wu F (2012) Towards new regionalism? Case study of changing regional governance in the Yangtze River Delta. Asia Pac Viewp 53(2):178–195
24. Li Y, Wu F (2013) The emergence of centrally initiated regional plan in China: a case study of Yangtze River Delta regional plan. Habitat Int 39:137–147
25. Li Y, Wu F, Hay I (2015) City-region integration policies and their incongruous outcomes: the case of Shantou-Chaozhou-Jieyang city-region in east Guangdong province, China. Habitat Int 46:214–222
26. Li Z, Xu J, Yeh AGO (2014) State rescaling and the making of city-regions in the Pearl River Delta, China. Environ Plan C 32(1):129–143
27. Luo X, Shen J (2009) A study on inter-city cooperation in the Yangtze River Delta region, China. Habitat Int 33(1):52–62
28. National Bureau of Statistics of China (2016) China city statistical yearbook. China Statistics Press, Beijing [中国国家统计局. 2016. 中国城市统计年鉴. 北京: 中国统计出版社]
29. Scott AJ et al (eds) (2001). Oxford University Press, Oxford
30. Smith N (1984) Uneven development: nature, capital, and the production of space. Blackwell, Oxford
31. Soja EW (1980) The socio-spatial dialectic. Ann Assoc Am Geogr 70(2):207–225
32. Soja EW (1989) Postmodern geographies: the reassertion of space in critical social theory. Verso, London
33. Soja EW (1996) Third space: journeys to Los Angeles and other real-and-imagined places. Blackwell, Oxford
34. Statistical Bureau of Anhui (2016) Anhui statistical yearbook. China Statistics Press, Beijing [安徽省统计局. 2016. 安徽统计年鉴. 北京: 中国统计出版社]
35. Statistical Bureau of Jiangsu (2016) Jiangsu statistical yearbook. China Statistics Press, Beijing [江苏省统计局. 2016. 江苏统计年鉴. 北京: 中国统计出版社]
36. Statistical Bureau of Shanghai (2016) Shanghai statistical yearbook. China Statistics Press, Beijing [上海市统计局. 2016. 上海统计年鉴. 北京: 中国统计出版社]
37. Statistical Bureau of Zhejiang (2016) Zhejiang statistical yearbook. China Statistics Press, Beijing [浙江省统计局. 2016. 浙江统计年鉴. 北京: 中国统计出版社]
38. Tomita K (1988) Geographical studies on structural changes in major metropolitan areas in Japan. Jpn J Hum Geogr 40(1):40–63
39. Wu F (2016) China's emergent city-region governance: a new form of state spatial selectivity through state-orchestrated rescaling. Int J Urban Reg Res 40(6):1134–1151
40. Yang D, Ye C, Wang X, Lu D, Xu J, Yang H (2018) Global distribution and evolvement of urbanization and PM2.5 (1998–2015). Atmos Environ 182:171–178
41. Ye C, Chen M, Chen R, Guo Z (2014) Multi-scalar separations: land use and production of space of Xianlin, a university town in Nanjing, China. Habitat Int 42:264–272
42. Ye C, Chen M, Duan J, Yang D (2017) Uneven development, urbanization and production of space in the middle-scale region based on the case of Jiangsu province, China. Habitat Int 66:106–116
43. Ye C, Zhu J, Li S, Yang S, Chen M (2019) Assessment and analysis of regional economic collaborative development within an urban agglomeration: Yangtze River Delta as a case. Habitat Int 83:20–29
44. Ye L (2014) State-led metropolitan governance in China: making integrated city regions. Cities 41:200–208

45. Yue W, Chen Y, Zhang Q, Liu Y (2019) Spatial explicit assessment of urban vitality using multi-source data: a case of Shanghai, China. Sustainability 11(3):638
46. Zhang J, Wu F (2006) China's changing economic governance: administrative annexation and the reorganization of local governments in the Yangtze River Delta. Reg Stud 40(1):3–21

Chapter 3
Production of Space in the Middle-Scale Region in China

Abstract The space in the process of urbanization is undergoing a great transition from physical space to social space. The relations between society and space become more important and complicated than ever. As a critical social theory, production of space means that urbanization has been reshaped by social factors or forces like capital, power and class, so that the urban space finally becomes their production and process. Based on socio-spatial dialectic, the main methodology of spatial production, urban space (re)shapes social relations and processes. The different spatial scales have and continuously produce the different social relations. In the researches about production of space, little work has been done on the index system to assess the extent of production of space and to analyze middle-scale region. This article designs a set of simple index system to reflect the spatial influences of capital, power as well as class, and chooses Jiangsu Province as a typical case because of its rapid and differential urbanization to indicate the process of spatial production from 2000 to 2015. In this index system, capital is represented by fixed-asset investment, real estate investment and foreign direct investment; power is represented by the index of the constructed urban land area; the index reflecting class is the urban rural income ratio. Based on the analysis of these changing indexes, this paper finds that urbanization in Jiangsu is hybrid process: the forces from capital and power greatly contribute to rapid urbanization and high urbanization level, however, the gaps among the three sub-regions in the province and rural–urban income inequality have not decreased accordingly. Conversely, these gaps enlarge in some periods. This kind of urbanization pattern is characterized by large-scale spatial expansion, and is driven by capital and power, but there are many latent social risks and spatial inequality. In the process of spatial production, space and society interact, entangle and (re)shape the pattern of urbanization in the end.

This chapter is based on [*Habitat International*, Ye, C., Chen, M., Duan, J., & Yang, D. (2017). Uneven development, urbanization and production of space in the middle-scale region based on the case of Jiangsu province, China. *Habitat International*, 66, 106–116].

3.1 Introduction

Since 2000, owing to unprecedented rapid and large-scale urbanization in human history, urbanization in the world, especially in some developing countries like China, has drawn more attention and has been studied by many scholars from the different areas [1, 7, 9, 15, 35, 8, 51, 53]. The kind of urbanization has rendered the spatial issue more remarkable than ever because producing and exploiting space to the maximum extent is the most important characteristic of urbanization, particularly in the capitalist time and world [21, 27, 34]. This process of urbanization is actually a process of production of space whereas production of space and urbanization are just two sides of a coin.

From "production in space" to "production of space", epistemology or philosophy on space has been greatly, even radically changed since 1970s, and space has become a key word and hot topic in social sciences, humanities and other fields [12–14, 19, 30, 47]. This kind of change has influenced society and everyday life, including urbanization. Space is no longer regarded as a dead, unchanging and empty object, or a physical or abstract factor or container without social connections or relations in the classic philosophers' views, including Newton, Kant and Marx, on the contrary, space is a production of social life, but more importantly, it is a subject which is never-ending, changing, (re)creating and (re)shaping society [12, 14, 30]. Space implies interrelations, interactions and multiplicity, and most importantly, it is socially constructed and "always in the process of being made" [38]. It is these new values and assessments on space that produce such a long-term tide studying production of space over the last four decades.

Since the masterpiece Production of Space (French edition published in the 1970s, English edition in 1991) written by Henri Lefebvre was published, Lefebvre and his followers have continuously developed this theory to integrate production of space, uneven geographical development into urban issues or urbanization research [6, 20, 47]. According to the critique and sublation to Lefebvre's thought, scholars developed some new frameworks in order to focus on key issues or methodologies [16, 41, 52]. Planetary urbanization characterized by production of space, is put forward, which emphasized on the intertwined temporal and spatial scales and their impacts on global urbanization [3, 4]. There is a dialectic, multi-scalar relation between urbanization and production of space. On the one hand, rapid urbanization not only changes the physical space including built environment, land use and landscapes and so on, but also reshapes the imaginative and mental space. For example, from the countryside to the city, spatial transformation often makes it difficult for immigrants to build up a belonging feeling and identity like their home in the rural areas. On the other hand, space itself is an important driving force and producer to produce and push urbanization. It is inconceivable that urbanization lacks a spatial base or support.

There are many kinds of scales between urbanization and production of space [2]. These different social, spatial and temporal scales interact and intertwine [5, 43, 54]. As for space, it can generally be divided into some scales, from place, city, region to country and world, according to spatial size and spatial level, or three types as macro, meso and micro [17]. There are both differences and connections between these scales which often lead to scale up and down. Many scholars have studied macro-scale urbanization and production of space, which focus on capital circulation, spatial fix and uneven geographical development using the methodology of political economic analysis based on Marx's theories [20, 22, 26, 42, 45, 49]. On micro-scale production of space, the scholars have carried out some frameworks to integrate land use, spatial production and policies, and summarize the mechanism of production of space based on the case of university town [33, 54]. On the city level, Paris, Los Angeles, Baltimore and Shanghai, as the typical urbanizing cases in the world, are often adopted to explain the process of production of space mainly based on the approaches of art and text interpretation [23, 25, 28, 48].

Compared to lots of works concerning about micro–macro-scale and city-level researches, little work has been done about spatial production of middle-scale region, especially the region like a province. Besides, the theory of production of space is very abstract and ambiguous, so more importantly, it is a key but difficult issue how to evaluate and judge the extent and degree of production of space. In order to deal with the two problems, this paper tries to design a simple set of indexes to assess regional production of space based on a typical case of Jiangsu, a rapidly urbanizing province in China.

3.2 Methodology: The Evaluation Indexes on Production of Space

Uneven development is an important issue in economics, which reflects the complicated relations between market, trade and state or regional development [18, 29]. It has been a hot topic in geography, especially in Marxist geography since 1970s [17, 44]. The basic attitude to capitalism is radically different facing uneven development between economists and Marxist geographers. The former doesn't deny capitalism and often thinks uneven development as the result of marketing force and free trade, but the latter regards uneven development as the outcome of capitalist system or Neoliberalism. Therefore, Marxist geographers are often critical and radical, and intend to uncover the driving force of capitalism. Production of space is the best tool or theory for this work.

The theory of production of space is basically different from the old theories about space. It integrated the past theories about space into the new temporal and social-spatial context, together with capitalism, finally developed a new kind of theory. According to this new theory, space is a tool, a backdrop and a product in the whole process of social (space) production. As [11] clearly summarized:

> Space appears at every stage in the production of social reality. It is the context for production (everything has to happen in space); it is a tool in production (we use space to produce particular forms of social relation), and it is a product (capitalism produces its own spaces through processes such as uneven development). Space is suddenly everywhere and appears to have considerably more theoretical power than it did in spatial science.

As a critical social or urban theory, "production of space generally means that the urban landscapes and spatial structures have been reshaped by political, economic, and social factors, mainly including capital, power, and class, so that the urban space finally becomes their production and process" [54]. This concept extends the core idea of production of space, "(social) space is (social) production" [30], and here "social" is a concept in a broad sense, including political, economic, (narrow-sense) social and cultural and other factors or human behaviors. "Social processes produce scales and scales affecting the operation of social processes. Social processes and space—and hence scales—mutually intersect, constitute, and rebound upon one another in an inseparable chain of determinations" [17]. This is so-called socio-spatial dialectic emphasizing dialectic interactions between time, space, and society [46, 47]. These scales can be further subdivided. According to this methodology, "social" concept can be mainly divided into three parts: political, economic, and (narrow-sense) social, which correspond to three main factors: power, capital, and class respectively [54]. Politics is a process that all kinds of power struggle and play. Capital is the most important factor for economic actions, especially in the capitalist system, and it flows over different areas and makes uneven space [10, 20, 24]. Class in society occupies a very important position similar to that of capital in economy. Urbanization produces the different spaces for the different classes such as the living space between the low-income earners and high-income earners, which space is usually used to separate the poor from the rich [31, 32]. Totally speaking, urbanization is a process that these three factors and the urban space interact and intertwine.

Once the relatively important factors are selected and verified, evaluating them becomes a necessary step. However, most papers focus on explanation and analysis of production of space using examples or cases without index or only one index, but overlook to assess it with a set of indexes. In fact, it is not difficult to cope with this work if we understand production of space from above three main factors: capital, power and class, because statistically speaking, there are some corresponding indexes to these three factors.

An index system can be made up of these selected indexes (Table 3.1). In order to focus on the most important characteristic of production of space, we construct a simplified but clear index system based on the three main factors above. In this index system, capital is represented by two or three indexes: fixed-asset investment or real estate investment which is included in fixed-asset investment and often has the same changing tendency with the former, and foreign direct investment; the power is represented by the index of the built-up (constructed urban land) area because the governments at different levels, especially in the countries like China, are often the main driving force to make the constructed urban land extend and expand; the main index representing class is the rural–urban income ratio which can directly indicate the gap between the rural and the urban inhabitants. As for urbanization, we adopt the general index, urbanization ratio to reflect the level of urbanization.

The data of this paper is mainly obtained from the Jiangsu Statistical Yearbook and China City Statistical Yearbook [39, 40, 50]. Although some question the accuracy of the data of Chinese statistics, the official data is generally effective and reliable, especially for the analysis of the large-scale temporal and spatial scope. In the paper,

3.3 The Evolving Characteristics of Production of Space in the Process …

Table 3.1 Main indexes assessing production of space in the course of urbanization

Factors	Main indexes
Capital	FAI (fixed-asset investment), REI (real estate investment), FDI (foreign direct investment)
Power	The built-up area
Class	URIR (urban–rural income proportion)
Urbanization	Urbanization ratio

Jiangsu province from 2000 to 2015 is relatively large-scale region. Different from the data of micro-scale space like a town which can get data based on fieldwork, most of province-scale data are often from the Bureau of Statistics of China and statistical yearbook.

3.3 The Evolving Characteristics of Production of Space in the Process of Jiangsu's Urbanization

3.3.1 The Uneven Geographical and Developmental Base in Jiangsu

Jiangsu, with its capital in Nanjing, one of the leading provinces in industry, is an eastern coastal province bordering Zhejiang and Shanghai to the south in Mainland China. It is also the second smallest, but the first most populous and the most densely populated of the 22 provinces in Mainland China. Jiangsu's GDP (6509 billion yuan in 2014) ranks second highest in China's provinces, after Guangdong province with 6781 billion yuan in 2014 [39]. Jiangsu has a coastline of over 1000 km (620 mi) along the Yellow Sea, and the Yangtze River runs through the southern part of the province. The thirteen prefecture-level divisions of Jiangsu are subdivided into 97 county-level divisions including 55 districts, 21 county-level cities, and 21 counties [50].

Jiangsu is a province with an obvious uneven geographical development. Generally speaking, especially from the economic geographical, historical and cultural view, Jiangsu, which can be divided into three parts–southern part, middle part and northern part, traces back to Jiangnan Province ("south of the Yangtze River") in the seventeenth century, which originally included today's Jiangsu and Anhui. Before then, the northern and southern parts of Jiangsu had less connection with each other than they later did. Traditionally, South Jiangsu is referred to as the three more prosperous southern cities including Suzhou, Wuxi and Changzhou. Their culture (the "Jiangnan" culture shared much in common with Shanghai and Zhejiang) is more southern than the rest and is often referred to as the Wu. All the other parts of the province are dominated by the so-called "Jianghuai Culture", which means the culture in the area between the Yangtse River (Jiang) and Huaihe River (Huai), though not all of them lie within the district defined by the term. In history, the term North Jiangsu refers to the cities to the north of the Yangtze River. For cities of Nanjing and Zhenjiang, neither of the two terms (North Jiangsu and South Jiangsu) refers to them, because though they are to the south of the river, culturally they are still of the Jianghuai Region. Since about 1998, there is a new classification used frequently by the government and defined by economic means. It groups all the cities to the south of the Yangtse River as South Jiangsu, the cities of Yangzhou, Nantong and Taizhou as Middle Jiangsu, whereas all the rest as North Jiangsu.

Though the terms of classification are very complex, by cultural means only the very north cities of Xuzhou and Lianyungang are culturally north Chinese. All the rest areas of the province are culturally south, though the three South Jiangsu cities are more purely southern while the culture in other cities is more a transitional mixture dominated by the south.

Although Jiangsu is wealthier than most Chinese provinces, there also is a big cultural, and economic geographical gap between the south areas and the north areas. The GDP per capita of Southern Jiangsu was 125,002 yuan in 2015, but the number of Northern Jiangsu was only 55,127 yuan, which the former was over twice that of the latter. The GDP per capita of wealthiest city, Suzhou, was 136,702 yuan,

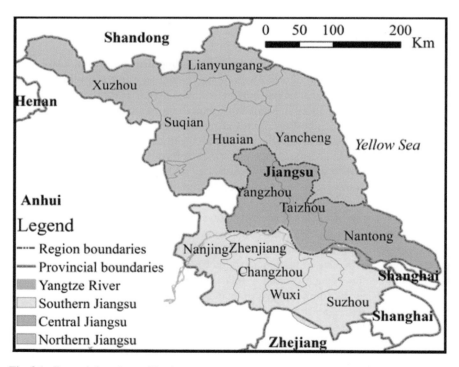

Fig. 3.1 Jiangsu's location and its three parts

3.3 The Evolving Characteristics of Production of Space in the Process … 37

and it was over three times of GDP per capita of Suqian city, 43,853 yuan in 2015. The uneven geographical and cultural backgrounds have led to uneven economic growth, and the increasing gap between the prosperous south areas and poorer north areas has become a historical problem in the development of Jiangsu, especially since the reform and opening policy in the 1980s. For example, [36, 37] states that China's economic reforms including the national policy "building a new socialist countryside" plays an important role in the process of rapid industrialization and urbanization in Southern Jiangsu, driving forces of farmland change, and. China's is an epoch-making countryside planning policy (see Fig. 3.1).

3.3.2 The Evolving Trend of Jiangsu's Urbanization Driven by Capital and Power

Great changes have been taking place rapidly in the process of urbanization in Jiangsu which can be demonstrated by the index of FAI (fixed assets investment), FDI (foreign direct investment), REI (real estate investment), URIR (urban–rural income proportion), urbanization ratio, and built-up area (Figs. 3.2, 3.3 and 3.4), FAI, REI and FDI show the influences of domestic and foreign capital and the evolving trends of these capital investment in Jiangsu. Totally, FAI, REI and urbanization ratio have basically the same changing trend. Regarding the strong relativity between urbanization and

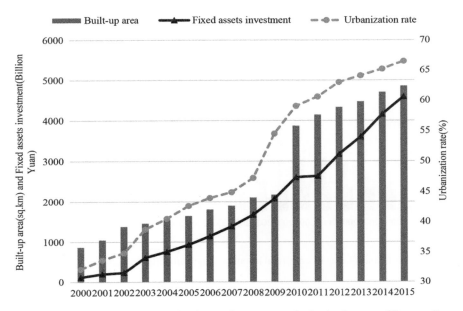

Fig. 3.2 Evolution of built-up area, fixed assets investment and urbanization rate of Jiangsu. *Data source* [39, 40, 50]

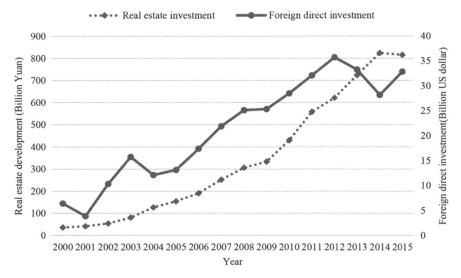

Fig. 3.3 Evolution of real estate development and foreign direct investment of Jiangsu. *Data source* [39, 40, 50]

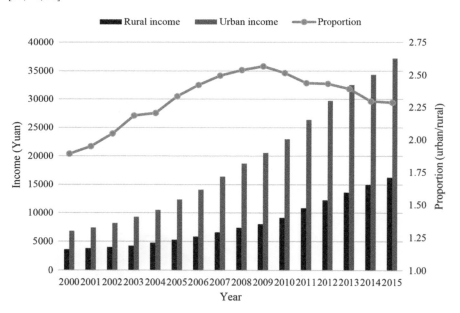

Fig. 3.4 Evolution of urban income, rural income and their proportion of Jiangsu. *Data source* [39, 40, 50]

domestic capital, it is the increasing FAI including REI that greatly and steadily contributes to urbanization. The growth rate of REI is much less than that of FAI. Therefore, the gap between the two indexes grows larger and larger from 2000 to 2015. Although FDI has increased in exponential growth generally and shared the same increasing trend with FAI, REI and urbanization ratio, there are some fluctuations. Even, FDI has declined in certain years, such as 2001, 2004, 2013, and 2014. This shows FDI has a strong but unstable effect on urbanization.

As for the index reflecting the influence of power, the built-up area of Jiangsu has been greatly expanding overall from 871 km^2 in 2000 to 4857 km^2 in 2015. And it has almost the same trajectory with the rapid urbanization. The year 2009 is a watershed dividing the different growth rate. During 2000–2009, the growth rate of built-up area was relatively low. It showed much higher growth rate after 2009 than before. There is an obvious and big leap of the built-up area from 2009 to 2010 which increases almost 1697 km^2 only one year. In fact, it is always a regular and important way for provincial and local governments to expand urban land as large as possible in order to realize rapid and high urbanization level. The ownership of land belongs to the whole state and local community. Therefore, the governments often expropriate the rural land or develop the other land into urban or industrial and business land in a very low price or freely. And then, redevelop these lands according to FAI and REI, and sell the constructed urban land for residents or developers in a high price. After China entered into WTO in 2001 and became opener than ever, FDI played a more important role in this process of urbanization. So, it is the collusion of power and capital that consistently produce the unprecedented space, which rapid urbanization need. So, it is not astonishing that FAI and REI, Built-up area, urbanization have shared very high synchronicity from 2000 to 2015.

Generally speaking, the rural–urban gap in Jiangsu is smaller than the national average level, which the URIR of the former is over 2 times, but the latter is over 3 times. However, the totally expanding rural–urban income disparity is not in accordance with the second wealthiest province in China. A clear differentiation or alienation between the rural and the urban in Jiangsu exists because uneven geographical development, resources and welfare system are distributed in an off-balance manner between the rural and the urban (Fig. 3.4). Urban-biased policies are the main cause for it. In the 1950s, the urban welfare system based on "hukou" institution, or household registration system, was established in order to supply various social services including the low-price provisions, housing, medical care, education and pensions for all urban residents. Conversely, rural welfare seldom improves. These urban-biased policies haven't been changed in the long term, so the urban–rural gap has gradually expanded, although reform and open-door policies have been applied for over 30 years. In addition, the coastal cities and counties often get more foreign and domestic investment but the rural areas are often overlooked, if not always. The rural areas are difficult to absorb the investment and capital because it is marginalized by the policymakers and developers. As a result, with the rapid development of urbanization, the urban–rural gap is widening stage by stage.

With the rapid and increasing urbanization ratio, the URIR continuously rises up from 1.89 of 2000 to 2.57 of 2009, first decreased and then increased, dividing at the point of year 2009. Before 2009, it was also divided into two parts. In the first stage, the URIR was relatively low because the urban and the rural both shared low-income levels. In the second stage, the index of URIR was relatively high because the income growth in the urban area grew faster than the rural area which led to a bigger gap between the urban income and the rural income. Thus, there are actually three stages of the change of URIR in Jiangsu from year 2000 to year 2015, which are characterized by balance at lower level, imbalance at higher level, and relative balance at higher level respectively. Overall, the urbanization ratio of Jiangsu has kept increasing from 2000 to 2015, of which the growth range reached its peak in 2010. The growth rate has been relatively steady and mild in other years.

3.3.3 The Constantly Growing City-Scale Inequity: Urbanization, Investment and the Rural–Urban Gap

The urbanization ratio of all cities in Jiangsu has increased significantly since 2000. Nanjing has been the top city while Suqian has been at the lowest level according to their urbanization ratio (Fig. 3.5). The gap of urbanization ratio among cities has narrowed down gradually. The difference range narrowed from 0.34 in 2000 (Nanjing as the maximum 0.568, Huaian as the minimum 0.229) to 0.27 in 2015 (Nanjing as the maximum 0.809, Suqian as the minimum 0.537), about 1.2 times. The spatial pattern shows evident regional differences between South Jiangsu, Middle Jiangsu, and North Jiangsu. The cities in South Jiangsu have the highest urbanization ratio followed by cities in Middle Jiangsu while cities in North Jiangsu have the relatively lowest urbanization ratio. The urbanization ratio of Nanjing, Wuxi, Suzhou, Changzhou, and Zhenjiang, as the leading cities in Jiangsu, is much higher than that of the other cities. The urbanization ratio of Suqian, Huaian, Lianyungang, Yancheng, and Xuzhou in North Jiangsu has been increasing at a fast speed but still relatively low. The changes and growth of urbanization ratio can be attributed to the changes in FDI, FAI, REI, URIR and built-up areas, which are going to be analyzed in the following text.

The index of FAI (fixed assets investment) is used to show the influence of capital on the urbanization process of Jiangsu. From 2000 to 2015, the amount of FAI in almost each city has been growing gradually (Fig. 3.6). However, there are evident differences between different parts of Jiangsu. South Jiangsu has the highest FAI followed by Middle Jiangsu then North Jiangsu. During this stage, Nanjing has been the leading city in the amount of FAI while the city of Suqian has generally remained at the bottom. Moreover, the gap of FAI between cities has been increasing during this time, ranging from 18 billion Yuan in 2000 (with Nanjing as the maximum 20, and Suqian as the minimum 2) to 359 billion Yuan in 2015 (with Nanjing as the maximum 543, and Suqian as the minimum 184), which is about 20 times. The amount of FAI in Xuzhou, which has been relatively high among the five cities in North Jiangsu, ranges from 10 to 427 billion Yuan in 2000–2015, while Suqian ranges from 2 to 184

3.3 The Evolving Characteristics of Production of Space in the Process … 41

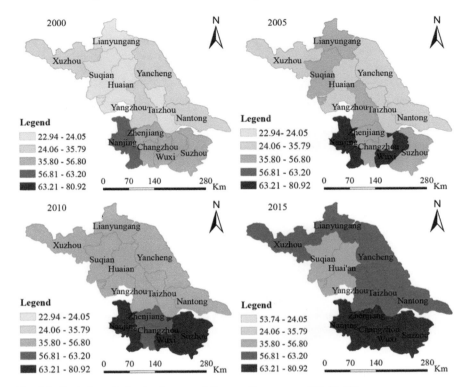

Fig. 3.5 Evolution of urbanization rate of Jiangsu. *Data source* [39, 40, 50]

billion Yuan. This fact indicates that even cities in the same part of Jiangsu such as North Jiangsu actually have huge differences in their capital investment, and these differences lead to their urbanization ratio difference shown (Fig. 3.5).

The index of FDI (foreign direct investment) can directly indicate the influence of foreign capital on the process of urbanization of Jiangsu. According to the spatial evolution of the FDI in Jiangsu from 2000 to 2015 (Fig. 3.7), there exists a big and obvious gap of foreign direct investment between South Jiangsu and North Jiangsu, though, the FDI of all cities in the whole province is constantly rising up in this stage. The cities of South Jiangsu have played a leading role since 2000. Mainly due to its superior geographical location, Suzhou, which is very close to Shanghai, has always been the top city in the amount of FDI. In 2000, the FDI of Suzhou was 1027 million dollars which increased to 6000 million dollars in 2015. In 2000, the city with the lowest FDI of 4 million dollars is Suqian. Its FDI reaches 298 million dollars in 2015. The difference on FDI between Suzhou and Suqian has narrowed from 256 to 20 times. However, there is actually a decrease of FDI from 8535 to 6000 million dollars from 2013 to 2015 in Suzhou. In 2015, Taizhou and Nantong also rank among those with the most amount of FDI after Suzhou, Nanjing, Wuxi, and Changzhou. The cities in North Jiangsu, such as Suqian, Huaian, and Yancheng, have a relatively lower FDI

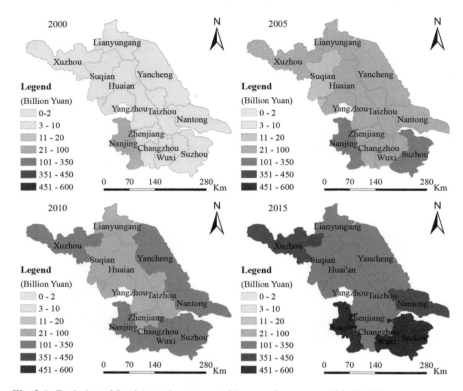

Fig. 3.6 Evolution of fixed assets investment of Jiangsu. *Data source* [39, 40, 50]

than South Jiangsu. This spatial distribution and evolvement of FDI have generally aligned with the changing trends of urbanization, and have basically not changed ever since 2000. There have been more and more industries and factories investing in Jiangsu, which helps break the situation that Hong Kong, Macau, and Taiwan were leading regions in the investment volume in the past. The United States of America and the developed countries in Europe have also increased their investment in Jiangsu since 2001 when China entered into WTO. The foreign investment has been infiltrating with increasing speed into service industries such as commerce, securities, and insurance banks and other industries through the entry point of manufacturing industry. The uneven distribution and growth of FDI in the three parts of Jiangsu are one of the main causes of the increasing gap between economic development in South Jiangsu, Middle Jiangsu and North Jiangsu (Fig. 3.8).

The index of built-up areas demonstrates the influence of power in the process of urbanization in Jiangsu. The built-up area has been rapidly expanding in a large scale in almost each city of Jiangsu every year. The regional and spatial differences are mainly shown in two aspects. First, the South Jiangsu shares a largest part of the built-up area followed by Middle Jiangsu then North Jiangsu. Second, the differences between cities are increasing during year 2000–2015. The difference range increased

3.3 The Evolving Characteristics of Production of Space in the Process … 43

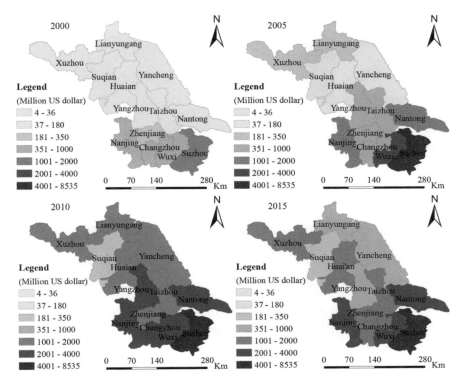

Fig. 3.7 Evolution of foreign direct investment of Jiangsu. *Data source* [39, 40, 50]

from 179 km^2 in 2000 (Nanjing as the maximum 201 km^2 is almost 10 times of Suqian as the minimum 22 km^2) to 552 km^2 in 2015 (Nanjing as the maximum 755 km^2, Taizhou as the minimum 203 km^2, which the former is over 3 times of the latter). In some cities like Suzhou, the changing trend of their built-up areas is similar to that of Jiangsu overall; however, the cities like Suqian have been playing an astonishing role in increasing their built-up areas, which increase from 22 km^2 in 2000 to 210 km^2 in 2015, an increase almost ten times as large as before. Nanjing has been the top city as for its expanding built-up area, which increased by 554 km^2 from 201 km^2 in 2000 to 755 km^2 in 2015. The growth of built-up areas in Nanjing has been greatly influenced by its status as the capital of Jiangsu. As the political center of Jiangsu, Nanjing is more powerful and dominating than other cities, which helps ensure the priority of Nanjing, including the largest built-up area.

In China, the rural and the urban are not only two kinds of space, but also two main social classes, and the residents of them have a deep gap on infrastructures, welfare, medical services and income, etc. The index of URIR (urban–rural income proportion) is mainly used to illustrate the influence of class in the process of Jiangsu's urbanization. For Jiangsu as a whole, the average URIR is becoming smaller, which means that the income coordination between the urban and the rural

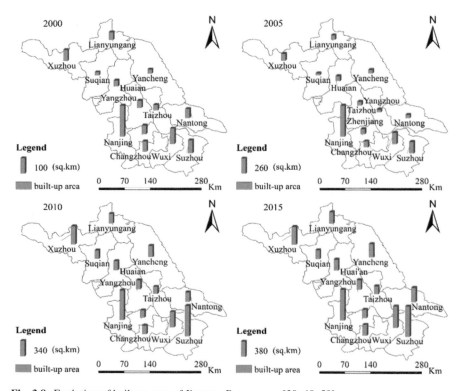

Fig. 3.8 Evolution of built-up area of Jiangsu. *Data source* [39, 40, 50]

has improved. However, there has been an evident difference between the different parts of Jiangsu. The differences between South Jiangsu, Middle Jiangsu, and North Jiangsu are smaller compared to those of capital and power. There are two regions with relatively low URIR, which are in South Jiangsu and North Jiangsu. In South Jiangsu, especially the region near Suzhou, Wuxi, and Changzhou, the URIR is low because of the high urban–rural coordination based on the well-developed economy. In North Jiangsu, especially in Suqian and Yancheng, the URIR is low because of the high urban–rural coordination based on the economic backwardness. The difference between cities has narrowed from 0.89 in 2000 (with Lianyungang as the maximum 2.49, and Suqian as the minimum 1.60) to 0.57 in 2015 (with Nanjing as the maximum 2.33, and Suqian as the minimum 1.76) (see Fig. 3.9).

3.4 Conclusions and Discussions

Space, society and time are interactive according to the principle of socio-spatial dialectic, and space of production or socio-spatial dialectic focuses on the interactions

3.4 Conclusions and Discussions

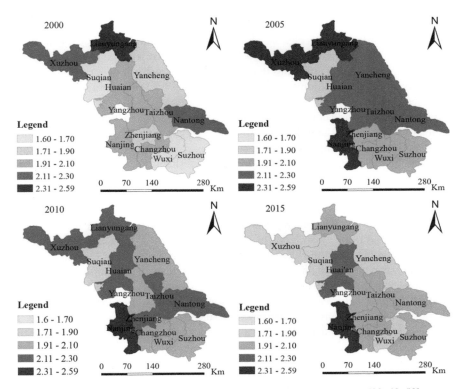

Fig. 3.9 Evolution of urban–rural income proportion of Jiangsu. *Data source* [39, 40, 50]

between time, space, and society of different levels or scales [54]. As [30] pointed out, social relations are spatial relations and vice versa, urbanization driven by capital dissimilates social (spatial) relations, and "every society—and hence every mode of production with its subvariants (i.e., all those societies which exemplify the general concept)—produces a space, its own space". Although the economic, social, and cultural and educational systems of China are very different from the western countries, they all produce their own spaces in some same or different ways, especially in the more and more globalized world. It is important to uncover and describe the characteristics of spatial convergences and differentiations in the process of urbanization, and discover the real driving forces or dynamics of urbanization in this capitalist world.

Uneven development is the key to understand production of space. Uneven development in Jiangsu province is firstly a result of the uneven geographical base and developmental context. Capital, power and their interaction produce significant effects on Jiangsu's urbanization. "Uneven development is the systematic geographical expression of the contradictions inherent in the very constitution and structure of capital" [45]. Therefore, here, geographical factors are not only physical conditions, built environments and locations, but also are a kind of complicated structure, system or process entangled with social factors and physical factors so that we can't divide them into the different parts clearly. There are no pure natural or physical factors, to some extent. The physical geographical conditions are (re)created or (re)shaped by some special societies. This means that the role of geography or geographical factors in the process of urbanization or production of space is becoming more different than ever. Geographical conditions or space are the base, context and dynamic of regional, national and even world development, and also are a process and finding. As space has been still redefined in the production of space, geography and geographical conditions, as the main driving force and result of spatial production, are redefined too. This kind of uneven geographical development is reshaping urbanization in different scales.

Scale is one of the most remarkable keywords or methodologies in the research of social space or production of space. Urbanization in China, together with production of space, in some middle-scale regions like a province as same as other scales, which can be explained from 3 interacting elements: capital, power and class. However, the middle-scale region is a little bit special. Especially, urbanization in Jiangsu is hybrid and greatly uneven. It is the main driving forces from uneven economic geographical conditions, capital and power that achieve the rapid and large-scale urbanization, but result in big gaps between the three sub-regions and rural–urban income in Jiangsu. Also, with the improvement of urbanization, these gaps are becoming widening. It seems that this process and result are almost inevitable if this pattern of urbanization can't be changed. This is one of the biggest risks in the process of China's urbanization.

With the coming of planetary urbanization, production of space takes place in many different scales and is full of any scales. Therefore, the middle-scale region is both special and common. On the one hand, this kind of region like Jiangsu has its special identity of place such as regional culture which is really different from the other region or place; on the other hand, urbanization and the development of Jiangsu are inevitably influenced by the larger region or state power and even globalization, meanwhile, Jiangsu also has an effect on the smaller spatial scale like the local. To some extent, this middle-scale is only defined from a traditional perspective, but the real region or space reflects or is always full of many scales. So, time and space can be divided into many kinds of scales and should be rethought from the different changing scales. It is high time to redefine or reorder urbanization and production of space.

References

1. Bloom DE, Canning D, Fink G (2008) Urbanization and the wealth of nations. Science 319(5864):772–775
2. Brenner N (2000) The urban question as a scale question: reflections on Henri Lefebvre, urban theory and the politics of scale. Int J Urban Reg Res 24(2):361–378
3. Brenner N (2013) Theses on urbanization. Publ Cult 25(1):85–114
4. Brenner N, Schmid C (2015) Towards a new epistemology of the urban? City 19(2–3):151–182

References

5. Cash D, Adger W, Berkes F, Garden P, Lebel L, Olsson P, Pritchard L, Young O (2006) Scale and cross-scale dynamics: governance and information in a multilevel world. Ecol Soc 11(2):8
6. Castells M (1977) The urban question (Sheridan A, Trans.). The MIT Press, Cambridge
7. Chan KW (2010) Fundamentals of China's urbanization and policy. China Review 10(1):63–93
8. Chen M, Liu W, Lu D (2016) Challenges and the way forward in China's new-type urbanization. Land Use Policy 55:334–339
9. Cohen B (2006) Urbanization in developing countries: current trends, future projections, and key challenges for sustainability. Technol Soc 28(1–2):63–80
10. Christophers B (2011) Revisiting the urbanization of capital. Ann Assoc Am Geogr 101(6):1347–1364
11. Cresswell T (2013) Geographic thought: a critical introduction. Wiley-Blackwell, Chichester
12. Foucault M (1977) Discipline and punishment. Tavistock, London
13. Foucault M (1980) Power/knowledge. Harvester, Brighton
14. Foucault M (1986) Of other spaces. Diacritics 16(1):22–27
15. Friedmann J (2006) Four theses in the study of China's urbanization. Int J Urban Reg Res 30(2):440–451
16. Gottdiener M (1985) The social production of urban space. University of Texas Press, Austin
17. Gregory D, Johnson R, Pratt G, Watts M, Whatmore S (2009) The dictionary of human geography, 5th edn. Blackwell, Oxford
18. Harris D (2008) Uneven development. In: Steven B, Lawrence B (eds) The new Palgrave dictionary of economics, 2nd edn. Palgrave Macmillan, New York
19. Harvey D (1973) Social justice and the city. Edward Arnold, London
20. Harvey D (1982) The limits to capital. Blackwell, Oxford
21. Harvey D (1985) The urbanization of capital. Blackwell, Oxford
22. Harvey D (1996) Justice, nature and the geography of difference. Blackwell, Cambridge
23. Harvey D (2000) Space of hope. Edinburgh University Press, Edinburgh
24. Harvey D (2001) Spaces of capital: towards a critical geography. Edinburgh University Press, Edinburgh
25. Harvey D (2003) Paris, capital of modernity. Routledge, New York
26. Harvey D (2010) The enigma of capital and the crises of capitalism. Profile Books, London
27. Harvey D (2012) Rebel cities: from the right to the city to the urban revolution. Verso, London
28. Huang Z (2004) The production of urban space in globalizing Shanghai. Taiwan Soc Res Q 53:61–83
29. Krugman P (1981) Trade, accumulation, and uneven development. J Dev Econ 8(2):149–161
30. Lefebvre H (1991) The production of space (Smith N, Trans.). Blackwell, Oxford
31. Lefebvre H (1996) Writings on cities. Blackwell, Oxford
32. Lefebvre H (2003) The urban revolution. University of Minnesota Press, Minneapolis
33. Li Z, Li X, Wang L (2014) Speculative urbanism and the making of university towns in China: a case of Guangzhou university town. Habitat Int 44:422–431
34. Lin GCS (2007) Chinese urbanism in question: state, society, and the reproduction of urban spaces. Urban Geogr 28(1):7–29
35. Long H, Li Y, Liu Y, Woods M, Zou J (2012) Accelerated restructuring in rural China fueled by 'increasing vs. decreasing balance' land-use policy for dealing with hollowed villages. Land Use Policy 29(1):11–22
36. Long H, Liu Y, Wu X, Dong G (2009) Spatio-temporal dynamic patterns of farmland and rural settlements in Su-Xi-Chang Region: implications for building a new countryside in coastal China. Land Use Policy 26(2):322–333
37. Long H, Tang G, Li X, Heilig GK (2007) Socio-economic driving forces of land-use change in Kunshan, the Yangtze River delta economic area of China. J Environ Manage 83(3):351–364
38. Massey D (2005) For space. The Sage, London
39. National Bureau of Statistics of China (2016) China statistical yearbook. China Statistical Press, Beijing [中国国家统计局. 2016. 中国统计年鉴. 北京: 中国统计出版社]
40. National Bureau of Statistics of China (2016) China city statistical yearbook 2000–2016. China Statistical Press, Beijing [中国国家统计局. 2016. 中国城市统计年鉴(2000–2016). 北京: 中国统计出版社]

41. Olds K (1995) Globalization and the production of new urban spaces: Pacific Rim megaprojects in the late 20th century. Environ Plan A 27(11):1713–1743
42. Quaini M (1982) Geography and Marxism. Blackwell, Oxford
43. Shen J (2005) Space, scale and the state: reorganizing urban space in China. In: Ma L, Wu F (eds) Restructuring the Chinese city: changing society, economy and space. Routledge, London
44. Smith N (1982) Gentrification and uneven development. Econ Geogr 58(2):139–155
45. Smith N (1984) Uneven development: nature, capital, and the production of space. Blackwell, Oxford
46. Soja EW (1980) The socio-spatial dialectic. Ann Assoc Am Geogr 70(2):207–225
47. Soja EW (1989) Postmodern geographies: the reassertion of space in critical social theory. Verso, London
48. Soja EW (1996) Third space: journeys to Los Angeles and other real-and-imagined places. Blackwell, Oxford
49. Soja EW (2010) Seeking spatial justice. University of Minnesota Press, Minneapolis
50. Statistical Bureau of Jiangsu (2016) Jiangsu statistical yearbook 2000–2016. China Statistics Press, Beijing [江苏省统计局. 2016. 江苏统计年鉴 (2000–2016). 北京: 中国统计出版社]
51. United Nations Department of Economic and Social Affairs (2012) World urbanization prospects: the 2011 revision. United Nations, New York
52. Unwin T (2000) A waste of space? Towards a critique of the social production of space. Trans Inst Br Geogr 25(1):11–29
53. Wu F (2016) Emerging Chinese cities: implications for global urban studies. Prof Geogr 68(2):338–348
54. Ye C, Chen M, Chen R, Guo Z (2014) Multi-scalar separations: land use and production of space in Xianlin, a university town in Nanjing, China. Habitat Int 42(2):264–272

Chapter 4
Spatial Production and Spatial Dialectic of New Urban Districts in China

Abstract New Urban Districts (NUDs) are the important spatial carriers to promote urban expansion or transformation. Since the 1990s, they have been playing a more and more crucial role in China's urbanization. For NUDs in the strict sense we found that: 96% to the east of Hu Line; 56% within the municipal districts; 64% within 36 km from their every city center and below the area of 423 km^2. The regional distribution follows significant spatial difference as "Eastern Region (50%)—Central Region (42%)—Western Region (8%)", and the provinces with the largest number of NUDs are Guangdong, Henan, Zhejiang, Liaoning, and Jiangsu. Furthermore, their interesting constructed process highlights the typical characteristics of spatial production and spatial dialectic. This paper uses the theory of production of space, and discovers that the growth of NUDs is a rapid ternary dialectical process of spatial production: "representations of space" is guided by the top-down governmental power; "spatial practice" is reflected in the hierarchical and regional difference of spatial elements, such as the type, pattern, distance and area of NUD; "spaces of representation" embodies the tension between governmental power and urban development rights, as well as the countermeasure mechanism. The extensibility of spatiotemporal sequences ensures the unity and continuity of spatial(re)production of NUDs. However, this is also facing a series of challenges like the management coordination of administrative division and the increasing unbalanced or inadequate development. Thus, critically rethinking the evolution of NUD is the key basis for achieving sustainable urban renewal and regional orderly development in the new era.

This chapter is based on [*Journal of Geographical Sciences*, Zhuang, L., Ye, C., & Hu, S. (2019). Spatial production and spatial dialectic: evidence from the New Urban Districts in China. *Journal of Geographical Sciences*, 29(12), 1981–1998].

4.1 New Urban Districts, Urbanization and Spatial Production

NUD is a very important phenomenon and concept in China. With the evolving urbanization, its development is to be more and more rapid and widespread. The concept of NUD can be traced back to the "garden city" which is essentially a combination of city and township [22]. In view of the ideas of sustainable and composite development, it has gradually become an inevitable choice for many countries or regions. Governments usually have a large number of institutional, planning, and policy tools that are key factors influencing NUDs [26]. In these new development zones, the transformation of land use in the suburbs often directly leads to cities' spatial expansion. However, because of the "mismatch" between conceptual goals and planning practices [47], NUDs also encounter various setbacks, such as socio-spatial segregation, excessive urbanization, environmental degradation and other unsustainable development issues [1, 13, 25, 43]. As the ultimate decision maker and participant of NUD establishment, government plays a powerful role in the expansion of urban emerging spaces [34]. By May 2016, the number of NUDs in China in the broadest sense has exceeded 3500 [12]. But the terms "new town" and "new district" are commonly mixed, so it is urgent to define the NUD in a specific sense through the definition of indicators as the main target for dialysis the development logic of China's urbanization. Obviously, large-scale and rapid traditional urbanization in China has put multiple pressures on the ecological environment, infrastructure and natural resources [7]. In future, the national strategy of new-type urbanization should emphasize the urban development appeals of people-oriented and urban–rural coordination [5, 6, 33]. In which, NUD will be a vital spatial carrier and support for the urban agglomeration planning in the new-type urbanization [58]. Since the 1990s, major cities have established district-or-county-level and even higher-level NUDs what have spread to 30 provinces in China Mainland except Beijing [60]. This growth process is full of ups and downs. Despite the emergence of high-level warnings and normative constraints in recent years, the intensive establishment of state-level NUDs remarks the importance of the central government in shaping urban transformation through territorial reforms [24, 35]. And the urban expansion of municipal districts is particularly prominent. Therefore, the spatial production of NUDs has developed into an unavoidable phenomenon and problem.

Research on the relationship between NUD and urbanization is receiving more and more attention and progress. The policy guidance for regional population mobility has reconstructed the urban system and spatial pattern [32], so China's social informality has been created by the "urban–rural dual structure" [48, 53]. By improving the quantity and quality of small and medium-sized cities, it will effectively eliminate the dual-track structure of urban–rural development and enhance the synergistic development capacities [31, 55]. In this context, NUDs came into being and became one main trend of city development in China, which also spawned new urban problems [44]. The domestic research topics on NUD have gone through three major stages: foreign theories, development zone, and comprehensive NUD. Of course,

4.1 New Urban Districts, Urbanization and Spatial Production

some studies have already touched on the relationship between spatial production and NUD: spatial production is an important way of capital proliferation in a specific period [2], and NUD is a main spatial carrier for "spatial fix" of surplus capital during the process of capital accumulation. Consequently, the internal structural adjustment of capital accumulation cycle has brought different types, patterns and characteristics to NUDs. With introduction of theories such as scale reconstruction and spatial production, the state-level NUD is interpreted as a specific scale reconstruction tool under the central government decision-making [15]. Starting from the theory of spatial production, power and capital promote the new urbanization and also cause the alienation of urban spaces [36, 54]. Therefore, it is necessary to further understand the special logic of capitalization of space. China's urbanization has typical "endogenous" characteristic and is strictly influenced by administrative forces [3]. On the whole, the theoretical interpretation of spatial organization such as NUD is relatively lacking based on the combination of spatial production and administrative division [8, 11, 51]. Under the people-oriented concept of new-type urbanization, we need to reorganize and rethink the NUD and its transformation with power and capital in space.

In short, current researches on NUD mainly focus on its general form of spatial organization. In theory, it tends to explore the city-industry integration, functional orientation, management system and development model, and pay less attention to the characteristics of specific spatial unit [4, 38]. Therefore, the number and scale-expansion of NUDs need new theories to explain and respond. In the empirical aspect, most researches are based on the cases of some national or local new districts, and gradually pay attention to land operation, administrative governance and local state restructuring [30, 50], but lack of profound analysis on the evolutionary characteristics of nationwide NUDs and their internal logic [14, 29, 39, 49, 57]. Therefore, we must focus on this spatial organization of NUD, and through theoretical transformation to achieve critical thinking on the evolution of time and space pattern. The production of space is a critical theory for explaining the NUD. It is proposed by Lefebvre, whose core category is the trinity of space. Its methodology is "ternary dialectic of space", ethical appeal is justice, essential attribute is political, and dominant trends are urbanization and globalization [27, 28]. After him, Harvey and Smith have given a good inheritance to the theory [18, 20, 41]. In a word, the production of space refers to the process of interaction and influence between capital, power, class and space [52, 56]. NUD can be understood as a kind of "spatial fix" of capital, or an effective solution for continuous accumulation [19]. This highlights the characteristics of spatial production in which multiple elements playing each other.

We make a new interpretation of the evolution process of China's NUDs from the theoretical perspective of spatial production. NUD is an important practice and carrier of the new-type urbanization strategy. Along with rapid urbanization, although study on NUD has been carried out, the problems about process and mechanism of spatial production have not yet been clarified. This paper is based on the research idea of "model construction—positive analysis—logical abstraction": firstly, try to propose a conceptual model of spatial production in China's NUDs; secondly, analyze their spatiotemporal pattern and evolution path; finally, use the conceptual triad to reveal the logic of spatial production of NUD. We aim to provide a theoretical exploration and empirical path for the relationship between urbanization and spatial production.

4.2 Research Data and Methods

4.2.1 Research Object and Description

4.2.1.1 Data Sources

The data acquisition process of this study is as follows: According to concept definition, the policy texts from the governments' official websites of 31 provincial-level administrative regions and 334 prefecture-level ones in the mainland are analyzed, and 224 NUD samples from 159 cities are screened (Fig. 4.1). The data such as the time, area and organization are mainly from the official websites of 224 NUDs, as well as the statistical yearbooks of relevant cities, including some field research data. The deadline for data statistics is at the end of December 2016. Notably, time refers to the actual establishment time of NUD management organization, aiming at circumventing some of the planned or pending immature NUDs. In addition, the spatial distance of the management organization from its upper administrative center is derived from "Ranging Tool of Baidu Map" for the linear distance measurement of 224 NUDs and the 159 cities to which they belong. Again, the spatial coordinates are obtained from "Baidu Coordinate Picker" for geolocation of all NUDs' management organizations. The data of county-level administrative division is approved with reference to the "National Administrative Division Information Inquiring Platform", and the map of China's administrative divisions comes from "National Geomatics Center of China".

4.2.1.2 Concept Definition

NUD is a long-standing term, and the academic community has yet to reach a unanimous statement. We define it in a narrow sense as: an independent new urban area that integrates production and life and has a clear management scope, considerable management authority and special management system based on urban development demands; a solution to the function of city center or the expansion of urban strategy; a modern space unit that highlights the functional composite features of city. Meanwhile, sample selection depends on the following indicators: first, it has the corresponding independent regulatory body; second, it has an independent government or a dispatched agency established by a government at or above the prefecture level; and third, its management system specifications are at or above the county level; fourth, it

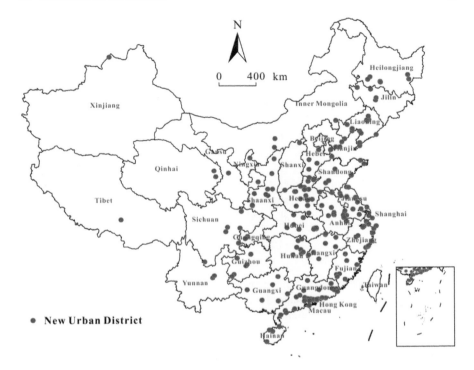

Fig. 4.1 Spatial distribution of 244 NUDs in China during 1993 to 2016

participates in the escrow, custody or direct management of grassroots administrative divisions such as villages and towns; fifth, it has the strategic positioning characteristics that reflect the city's composite function; sixth, its management organization (management committee or government) has an official website.

4.2.1.3 Research Area

In terms of time, the establishment time of NUD is mainly based on the actual listing of its management organization, and the statistical range is from 1993 to the end of 2016. In terms of space, it involves China's 31 provincial administrative regions that do not include Hong Kong, Macao and Taiwan. At regional level, the proportion of NUDs in the three Regions (Eastern, Central and Western) is 50%, 42% and 8% respectively. At provincial level, the provinces with a total number of NUDs accounting for more than 7% are Guangdong, Henan, Zhejiang, Liaoning, and Jiangsu. This paper divides the development of NUDs into three stages (Fig. 4.2). First is the Low-speed Growth Stage (LGS). It is characterized by a low growth and a slow rate, which is at a low-speed growth level before 2002 and a high volatility and low-speed growth at a later period. Next is the High-speed Expansion Stage

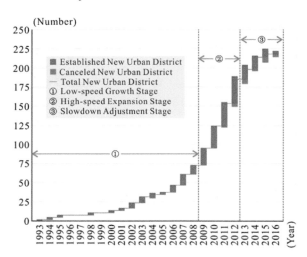

Fig. 4.2 The stage evolution of NUDs in China

(HES). The number achieved rapid growth and its rate is about 2 times higher than the highest value in the first stage. Its quantity also experienced a large expansion and the average annual increment is about 7 times in the first stage. And the annual number and increment are respectively the highest in history and are increasing year by year. Final is the Slowdown Adjustment Stage (SAS). The establishment of NUD quickly fell to the first stage level in a short period, and the annual average increment is less than 1/3 of HES, and only twice the LGS. At the same time, under the trend of continuous "going down", state-level ones have ushered in a rapid increase. Besides, the development of NUDs is not a process of "increasing growth". It has also undergone cancellation or transformation and experienced adjustments in terms of growth rate or establishment ways in recent years.

4.2.2 Research Framework and Method

4.2.2.1 Spatial Model

The construction of NUDs in the process of urbanization is closely related to the reconstruction of urban spaces in China. It is necessary to establish a spatial evolution model under the influence of power. Based on the theory of spatial production, this paper designs a model that reflects the construction of NUD is a three-dimensional, multi-factor interaction process of spatial production (Fig. 4.3): first, it has different functional characteristics from the traditional development zones; second, it is a spatial superposition process that follows the evolution of "development zone—NUD—administrative district"; third, it has different management systems and spatial setting types; fourth, its establishment and development is a bidirectional process of "top-down" and "bottom-up" [59]. Under the transformation demands of

4.2 Research Data and Methods

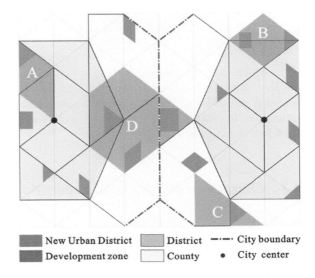

Fig. 4.3 Spatial model of NUDs in China

local traditional development models, diversified single functional zones need to be transformed into integrated ones. In the meantime, the traditional models and gradient strategies have exacerbated regional imbalances. At the national level, a new set of regional growth poles needs to be rapidly built through new spatial expansion models to promote the orderly development between regions. Thus, under the two-way effect of top-down and bottom-up, governments at different levels will set up new districts and give different development orientations. In addition, the pioneering development privileges have created more new spaces for local government-led urban expansion. Dominated by governmental power, NUD is a rapid process of spatial production that integrates the characteristics of "spatial evolution of complex functions", "centralized urban expansion", "multi-dimensional spatial superposition" and "strategic adjustment of regional development".

4.2.2.2 Method

The density estimation method of *Kernel* is applied significantly in the agglomeration analysis of spatial point data and has a good performance in revealing the distribution characteristics of point groups in NUDs. Points located within the search radius will be assigned different weight values, and the closer to the grid search center, the greater the weight value will be assigned. This method mainly achieves the spatial visualization of point distribution by transforming the discrete "points" into a continuous "point group" density map. Let the density value at p be $\lambda_h(p)$, then the estimated value formula is as follows:

$$\lambda_h(p) = \sum_{i=1}^{n} \frac{3}{\pi h^4}\left(1 - \frac{(p-p_i)^2}{h^2}\right)^2$$

In which, p is the location of NUD point to be estimated, p_i is the i-th point position in the circular search area with p as the center and h h as the radius. The selection value of h will directly affect the smoothness of point distribution density estimation. In order to make the results year-to-year comparable, this paper sets the same search radius and cell size.

4.3 Characteristics of Spatial Production in China's NUD

4.3.1 Representations of Space: Construction of Space by Power

The "representations of space" refers to the conceptualization of space, or the allocation of power, knowledge, and space. The materiality of mainstream social orders is contained in it and is thus justified [17]. In the practice of NUD, it is primarily represented by the abstract designs of planning ideas by governments and their appointed experts, which is actually the construction of power for space. Therefore, we explain the "representations of space" of NUD based on strategy formulation, policy design, administrative department, and dispatched agency.

4.3.1.1 Strategy Formulation and Policy Design

The development of NUD has always run through the construction of spatial powers between different levels of government. Their administrative means include the formulation and restriction of many strategic plans and policy texts, such government-led statutes imply the interests of different discourse rights. Initially, local governments generally promote the scale expansion of cities through industrial construction of various development zones. With the large number of single-function development zones, their functional appeals of social services and management have become increasingly prominent. It is urgent to take some more comprehensive measures to integrate functions in the development zones, especially to solve some institutional bottlenecks. However, the traditional regional development strategies have made regional differences increasingly serious. We must choose more comprehensive spatial carriers as new growth poles to drive urban or regional development. At the same time, local governments will begin large-scale NUD planning, and constantly pursued higher administrative levels in order to obtain more development privileges or policy supports. In this fierce declaration process, NUD with better foundations or more potentials will be awarded the title of "state-level". NUD has thus become the main mode of transformation of the development zone, and an important way for the urban spaces to carry out a new round of strategic expansion.

4.3.1.2 Administrative Department and Dispatched Agency

NUD reflects the government's top-down power allocation. Relying on the spatial layout of regional strategies, the State Council or local governments will selectively instruct specific areas to prepare for the construction of NUDs. If the approval process is successfully passed, it will be officially awarded the title of "NUD" at the corresponding level, and a strict top-down administrative management system will be formed. There are both the macro guidance of top management departments and the dispatched management of local governments' high-level configuration: first, NUD is usually approved by the government departments of higher authorities and is mainly supervised and guided by the National Development and Reform Commission (NDRC), and adopts a step-by-step periodic reporting mechanism; second, government departments (involved in land, environmental protection, and housing construction) shall comply with the development of NUDs in accordance with their establishment criteria and review procedures; third, local provinces (and autonomous regions or municipalities) need to set a coordination mechanism to coordinate and solve major problems in the development of NUDs. The main leaders of NUD management agencies are often directly or concurrently appointed by provincial or municipal cadres. In addition, the management committee is often used as the dispatched agency of provincial or municipal government and is responsible for economic development, urban construction or social management in accordance with the county-level or above organizational standards. At different stages, NUD will also adopt different administrative management systems according to its own actual conditions, which can be roughly divided into three types: the first type is Management Committee. As a dispatched agency of higher-level government, it largely exercises the management authorities of development and construction in the district, while its social affairs are basically the responsibility of the administrative district where it is located; the second one is Combination of Administrative Region with NUD. Namely, the management committee and its administrative district government implement a joint office of "a set of people and two brands"; the last type is Government. The NUD was approved by the State Council to establish a first-level government and was given full administrative authorities. Therefore, in the early stages and most of the current NUDs adopt the management committee type. In the more mature stages, it is necessary for NUD to break through the constraints of the original administrative district to undertake more construction, management or service functions, and then gradually establish a government to achieve the transition to an independent administrative district.

4.3.2 Spatial Practice: Shape of Space by Power

The "spatial practice" refers to the space–time routine and spatial structure (such as places and paths) of social life to achieve production and reproduction. It actually corresponds to the process by which power shapes the real space of physical form. There are four aspects: the production of space, the dominance and control of space, the occupation and use of space, the accessibility and distance of space [21]. Therefore, this paper selects the distribution pattern, combination type, center distance and area size as the four practical references of the "spatial practice".

4.3.2.1 Distribution Pattern and Combination Type

On the one hand, the evolution study of spatiotemporal patterns is conducive to revealing a series of phased processes and outcomes of NUDs, which are related to government behavior, land use, resource environment, social life and many other aspects. Through the *Kernel* analysis in the node years, it is found that NUDs have significant spatial differentiation characteristics (Fig. 4.4): (a) By 2004, NUDs were distributed evenly in the eastern and central provinces of the Yangtze and Yellow River basins, but it had not yet been established in South China; (b) By 2008, NUDs were gradually launched in many provinces in the east of the Hu Line, which were the most significant in the eastern coastal areas and had the highest density in the Yangtze River Delta (YRD). During the period, all the three provinces of South China began to set up NUDs in 2007; (c) By 2012, the NUDs distributed to the west of the Hu Line accounted for about 5% of the total. At this time, the spatial agglomeration characteristics of the national scale are particularly obvious, and there are high-density clusters such as "Jiangsu-Zhejiang-Shanghai-Anhui", and sub-clusters in Henan and Guangdong, as well as the largest spatial agglomeration belt running through the north and south; (d) The number of NUDs existed in 2016 is 27% higher than that in 2012, and the major trend of clusters and belts is further strengthened. Moreover, the number of wests of the Hu Line accounted for only 4%, and the regional differences in spatial distribution continued to widen.

On the other hand, to explore the combination relationship between NUD and municipal district(s) or related county(s) is easy to grasp the spatial evolution logic of NUD in the process of urbanization. The spatial arrangement in NUDs can be classified into four types (Fig. 4.3; Table 4.1). The combination of A-type is mainly based on single "district", which is often expressed as the "upgrading" of NUD within the city's jurisdiction and the gradual "expansion" of the administrative divisions such as withdrawing county (or county-level city) and to set district. The combination of B-type mainly includes: district + county, district + county-level city, district + county + county-level city, district + autonomous banner. And the combination of C-type is as follows: county, county-level city, county + county-level city, autonomous county. Besides, the D-type representing the regional cooperation between cities accounts

4.3 Characteristics of Spatial Production in China's NUD

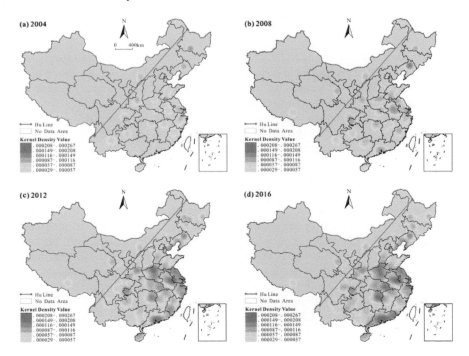

Fig. 4.4 Kernel analysis of spatial layout of NUDs in China during 1993–2016

for about 3%, and its combination ways mainly include: district + county + county-level city, district + county. It can be seen that the type evolution of "A-B-C-D" suggests the continuous breakthrough of NUDs at all levels in urban expansion, that is, gradually realize the co-construction from the city center to the outer suburbs and even adjacent cities in the spatial distance. Therefore, this also illustrates the inherent demands of the prefecture-level administrative region (or municipality) in NUD for urban expansion in the urbanization process.

Table 4.1 Types of spatial arrangement in NUDs

Type	Specific meaning	Proportion/%
A	Composed of single district(s) of the same prefecture-level city (or municipality)	56
B	Composed of district(s) and county(s) of the same prefecture-level city (or municipality)	20
C	Composed of a single county(s) of the same prefecture-level city (or municipality)	21
D	Composed of district(s) and county(s) of the different prefecture-level city (or municipality)	3

4.3.2.2 Center Distance and Area Size

Through the establishment of NUD, government has constructed a hierarchical practice mode of nested geographical distance and space scale. With the increasing planning scope of NUD, the location of its regulatory agency will gradually realize the extension of power allocation from municipal districts. Based on the mutual distance between the regulatory agency and its superior administrative center, this paper uses "Jenks" tool of ArcGIS10.2 to realize cluster analysis and spatial visualization of all distance data (Fig. 4.5, left): the NUDs of Grade-I account for nearly 60% of the total, and their spatial distribution in the central and western regions is relatively balanced and mainly concentrated in the eastern and central provinces between Yangtze and Yellow River. More than 96% of regulatory agencies are located within the municipal districts, having administrative convenience and efficiency. In which, several provincial capitals with only one district often locate its NUD outside its municipal district, while the regulatory agency is located within it. This tendency to "upgrade NUD to district" also symbolizes the centrality of management, so the inconsistency between NUD limits and its regulatory agency occurs within a short distance. The proportion of Grade-II is more than 1/4 and there are differences in the spatial distribution of the central and western regions, which are obviously concentrated in the Pearl River Delta (PRD), YRD and the Bohai Rim. Meanwhile, more than 1/2 of management agencies in the closer range are located within A-type NUDs, continuing a "point-to-face" consistency of the regulatory agency and NUD range within the city's municipal districts. Among them, 29% of NUDs' regulatory agencies are not located within the scope of municipal districts, so the NUDs outside the municipal districts have a certain space-consistency and management-independence. The regulatory agencies of Grade-III and Grade-IV show more obvious differences in spatial distribution. Grade-III is spatially distributed in PRD, YRD and the Beijing-Tianjin-Hebei Region on the eastern coast, while Grade-IV are scattered in Hebei, Liaoning, Hubei, Fujian, Hainan, and other provinces. These long-distance regulatory agencies often have the characteristics of "coastal trend" or "provincial boundary". The area of traditional new town in the world is mostly within the range of tens of square kilometers, while the one in China is generally larger, even exceeding the area of a county or district in the region [9, 23, 45]. As a result, NUDs with different area sizes and spatial distributions are significantly different (Table 4.2):

The Small-scale NUDs are spread over most provinces. Since the smallest size is common in the initial stage of urban expansion, they are more evenly distributed in the eastern, central and western regions. Most of the Medium-scale NUDs are distributed in the central and eastern coastal areas. Since the smaller size is often present the development stage of urban expansion, the distribution tends to shift to the central and eastern regions. The most important feature of Large-scale NUDs is that they are all located in the east of Hu Line. Since the large-scale planning is usually based on the relatively mature stage of city development, the distribution is more consistent with the regional urbanization level. The Massive-scale NUDs (including more than 3/5 state-level ones) are primarily distributed in the eastern coastal areas with high urbanization levels. With the emphasis on planning scope and regional

4.3 Characteristics of Spatial Production in China's NUD

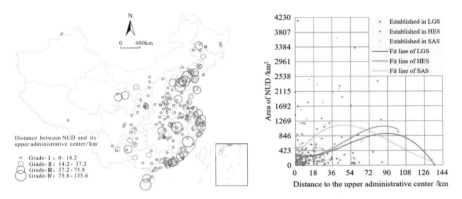

Fig. 4.5 The spatial distribution of distances between regulatory agency and upper administration center and the changing trend of spatial elements of NUDs in China during 1993–2016

Table 4.2 Area scale of NUDs

Scale	Feature	Area/km^2	Number ratio/%	Area ratio/%	Average area/km^2
Small	The smallest size, the largest number	0–160	47.8	8.7	77
Medium	The smaller size, the larger number	160–419	22.8	13.7	255
Large	The largest size, the smallest number	419–960	17.0	23.6	588
Massive	The largest size, the smallest number	960–4132	12.1	53.8	1887

strategy, the types of administrative districts spanned by Massive-scale NUD are also more complicated. In addition, it is found by a scatter plot of the phased distance and area data (Fig. 4.5, right): most of NUDs are concentrated within a distance of 18 km and an area of 423 km^2 or less, and it is gradually characterized by a more significant "closer-distance, larger-scale" planning.

4.3.3 Spaces of Representation: Reconstruction of Space by Power

The "spaces of representation" refers to the confront spaces or struggle places, which stem from the privacy and underside of social life, and the critical art of "questioning the practice of mainstream spaces and spatiality with imagination" [17]. In NUDs, this is externalized into the social relationship between bodies in our daily life, reconstructing the spaces with power through time and use. Therefore, we aim to interpret the "spaces of representation" from four aspects: power game, power appeal, development disorder, and planning early-warning.

4.3.3.1 Power Game and Right Appeal

Due to the lack of uniform standards and supervision mechanisms, the NUDs instructed or issued by prefecture-level cities occupy the largest proportion, and the power relationship within their spatial competition is more complicated. NUD fully embodies the spatial power game between the governments at all levels in urbanization. In this process, the government seat as a central place of power plays an important role. Since the development pressures of old districts and their adjacent counties, many cities will face the problem of urban district expansion and administrative center migration. Most of NUDs have become important transitional spaces for cities before the relocation of government seats. When the government offices successfully relocate, some NUDs will be withdrawn or converted, while other reserved ones will continue to assume responsibility for urban construction, and more than 13% involve the migration of prefecture-level or provincial-level administrative centers. In particular, some even moved the prefecture-level administrative center from municipal district to farther county. It shows that the local government is so powerful in exercising administrative means. At the same time, the construction of NUD has also highlighted the local struggles for urban development rights. First of all, the built-up area within the municipal districts is constantly saturated or the urban expansion is limited due to the "protection of resources and cultural relics". Therefore, the reconstruction or protection of old towns and the government transfer became one original intention of NUD establishment. Secondly, NUD is also the demand for the transformation and development of traditional resource-exhausted cities, and the western mountain cities are often faced with bottleneck constraints of linear cities due to geological terrain limitations. Again, the distribution of rivers, lakes and seas is another factor that affects the spatial layout of urban centers. As a spatial strategic form, NUD directly maps the future orientation of city development. In samples, NUDs with waterfront distribution are more than 1/2 and mostly named after the rivers, lakes and seas. Especially under a series of strategic guidance, the demands of waterfront development, cross-river development and basin co-development have been quite remarkable. For example, co-development can not only expand the main strength of immature areas, but also can learn from each other and achieve regional division of labor between regions.

4.3.3.2 Development Disorder and Planning Early-Warning

The surge in NUDs has stimulated and formed some developmental disorders. Firstly, in the long-term process of population-urbanization lags behind land-urbanization, the extension of urban spaces has emerged as a prominent phenomenon such as "ghost city", and similar problems even exist in the state-level NUDs such as Lanzhou New District. Now, more than 90% of the prefecture-level cities are planning NUDs and some of them have several times the total area of their respective built-up areas. Next, the NUDs also bring potential challenges and difficulties in administrative division work: many cities have promoted the transformation of NUDs into administrative districts by means of withdrawing counties or county-level cities, which will inevitably produce a strong demonstration effect. Moreover, the number of sub-level NUDs has also increased, and the NUD has gradually strengthened the custody and escrow of villages and towns, which have created new difficulties in administrative division management. Furthermore, the construction boom of NUD is still the key development orientation of various cities, and there are considerable NUDs that are under construction or have not yet established formal management agencies. More urban strategies or plans are also carrying out high-standard ideas for NUDs, especially in the "13th Five-Year Plan" of some provincial or prefecture-level cities. Among them, most plans have been put into action, and the goal is high and unrealistic has become a problem that cannot be ignored. Therefore, the establishment of NUDs in recent years has been subject to high-level warnings and is clearly bound in relevant documents. Planning early-warnings ensure that the problem of out-of-control is dealt with on the basis of scientific analysis such as data and facts. The approval of NUD mainly follows the processes such as "NDRC—Ministry of Housing and Urban–Rural Development (MOHURD)—Ministry of Natural Resources (MONR)": firstly, it strives to be included in the regional development plans by NDRC to form a concept of "NUD" and obtain corresponding policy supports; secondly, NUD will be planned and compiled for urban and rural construction by MOHURD; finally, the MONR needs to coordinate and weigh the indicators for construction land, which is also the key link for the legal approval of NUD. In addition to the above-mentioned norms, it also involves other factors such as urban hierarchy and government officials. In recent years, MONR has also tried to promote the implementation of the "national land planning control" system. That is to say, the NUD that conforms to the national planning is basically approved, otherwise it will be rejected. Therefore, planning early-warning should guarantee the binding effect of relevant normative documents and strictly control the approval of land use in NUDs.

4.4 Logics of Spatial Dialectic in China's NUD

To study the growth of China's NUDs from the perspective of spatial production, it should follow the dialectical and unified thinking of "time (process)—space (pattern)—society (mechanism)", and then establish a concise logic framework (Fig. 4.6). In which, the representations of space, spatial practice and spaces of representation are the core categories of Lefebvre's "ternary dialectic of space" [27]. There is a non-linear order between the three, but a mutual generation and construction relationship with multiple dialectic and tension. The "spaces of representation" and "representations of space" are antagonistic by "dominance space" and "opposition space" respectively, but the two are also intertwined. That is, the former helps the dominator to recognize the situation and position to seek the path of resistance, and the latter is the place where the dominant interests and resistance forces fight together and misappropriate [46]. Correspondingly, the "spatial practice" lies between the above two and makes them mutually interconnected and distinct. It is the space–time structure in which space can be produced and reproduced, which respectively supports the normal operation of "spaces of representation" and "representations of space". The "spatial practice" is a spatial process that promotes both production and reproduction with a certain continuity and cohesion [17, 37]. Therefore, this framework not only performs spatial interpretation based on society and history, but also interprets society and history based on space, and highlights the dialectical unity between society, space and time through the spiral evolution of the three categories.

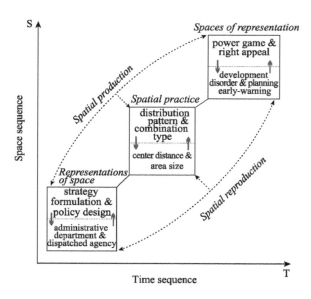

Fig. 4.6 The spatial production of NUDs and its ternary dialectic

4.4.1 *"Trinity": Category and Connotation*

The "representations of space" is mainly the conceptualized dominance space where various experts (and administrative agents and other agents or institutions) conduct norms and statutes through the symbolic system of language, discourse, text, and law. It is a spatial performance of the (political) power and ideology to achieve the concept of regularization [10, 37]. This category is embodied in the policy orientation and power allocation of NUDs at all levels of government, such as relying on scientists, planners, engineers to participate in strategy formulation, policy design to achieve textual planning for spaces, then through the administrative departments and their dispatched agencies further dominate the spatial control and benefit distribution in NUDs. In the process of developing NUD, government is not only the maker of space rules such as political texts and management systems, but also the biggest beneficiary of potential benefits such as land finance and social economy in space. It regards the NUD as a spatial tool to control, reconstruct and conquer the urban space. Thus, the government has the absolute right of discourse power. This is actually a top-down "discipline space" from the government perspective, which embodies the spatial representations of state orientation, government dominance and special policies in NUD model. This in turn allows officials, planners and developers to form meaningful abstract symbols (to guide spatial practice) [16].

The "spatial practice" is largely used as a "material form" process for producing social spaces. It is not only a medium for sensing behavioral activities and experiences in the external physical environment, but also a construction result including production and reproduction, concept and execution, conception and life. Furthermore, it is an important way to generate, use and perceive space [37, 40]. This category is specifically represented by the distribution pattern and spatial combination type from the NUD level, and further microscopically focuses on the two large spatial elements such as the size (of NUD) and the distance (between the management agency and its superior administrative center). It aims to examine the practical characteristics of NUD in the process of passive obedience to government spatial planning from different scales. The process of spatialization of governments' powers is also a process of impairing the rights to urban development. For example, the unbalanced spatial distribution and management radius of NUD is always accompanied by the prominent differences in urban development rights within and outside the region, which laterally reflects the imbalance and inadequacy of spatial development. This is actually a kind of perceptual production of space, corresponding to the material relationship between the new growth pole and the spatial superposition that is produced and used in the evolution of NUD. Consequently, the practice of NUDs is repeatedly organized, restricted and promoted by structure (pattern-type-distance-scale), and structure is not only the medium of practice but also the result or product of practice, so it is embedded in or embodied in the practice of NUDs. It eventually exists and plays a role in the form of memory traces [46].

The "spaces of representation" primarily refers to the imagination originating from the private secrets of social life, and is a space for questioning and criticizing mainstream spatial practice while giving its symbol and meaning. It is also a spatial field that combines realistic imagination with practical thinking to create an anti-dominant order [17, 42]. This category is embodied in the power game between local government departments at all levels and the struggle for development rights between local cities. The series of social effects generated by NUD under the "discipline" of power further stimulated the local struggle for urban rights. That is to say, the spatial planning of government powers is heterogeneous. Where the space rules fail, it is where the pursuit of production of "anti-discipline space". Therefore, the adjustment of administrative divisions and the planning of early-warning mechanisms have become the key means for government to carry out space adjustment and re-discipline. This is actually a kind of bottom-up production of "anti-discipline space", which is highlighted by the spatial struggles and counter-measures of spatial expansion, regional differences and functional integration of various development zones in the process of NUDs. This is in order to change the status of their subordinates or marginalization in the game [42].

4.4.2 Spatial Production and Reproduction

The interaction and dialectical unity of the above three categories is an important link and composition of the spatial production and reproduction process in NUDs. The reciprocal spiral advancement of "spatial triad" is the main way of spatial production.

The complementarity between strategic policies and departmental institutions consolidates the important rules or habits that are associated with governments at all levels in the urbanization system, and is a key tool for government power to guide spatial behaviors. Government-led model of NUD has its own fixed capital in terms of spiritual will, material practice, and social feedback. And the re-territorialization and re-spatialization of capital will increase the time–space complexity of unbalanced development. This can be a superior resource under government actions, or it can be a constraint to behaviors. Again, government's own design and acceptance of the disciplined approaches has shaped its spatial behaviors, and the differential rent-driven benefits driven by land finance have produced geographic differences in the intensity of capital investment. Therefore, spatial allocation of government powers fundamentally reshapes the combined mode of urban spaces. Especially, when such disciplinary texts and management designs are increasingly losing the relevance of spatial practice within local political system, the claims of urban development will become more prominent.

Unbalanced geographic development requires a combination of changing spatial scales and political behaviors, as well as the differences and recurrence of power behind different scales. The unbalanced and inadequate spatial and temporal pattern of NUD is being produced, shaped, adapted and reconstructed by the changing "political-economic" and "social-ecological" processes. The expansion of NUD size and the change of management distance highlight the important strategic significance of volatility and liquidity in the contemporary geospatial forms. Moreover, different actors and agents try to implement spatial planning behaviors by trans-administrative districts, thus realizing the upward shift and reconstruction of new urban districts at different levels such as "city-region-country".

NUD embodies the special policies pursued by specific powers in specific places. These power structures have obtained a lot of wealth interests and political discourse rights from spatial production. Meanwhile, NUD is facing development difficulties at different levels, scales and locations, such as limited capital, inadequate popularity, deficiency in management, debt risk, overcapacity, traffic congestion and ecological governance, environmental crisis and other urban issues. The tension between political powers and social rights is constantly being reshaped in the process of spatial production and reproduction in NUDs. Regional imbalances have created enormous resistance to governmental decisions that represent multiple sets of special interests. The established policies have obvious instability and imbalance. However, after the government regulates and intervenes in the spatial factors such as the number and size of NUDs, the imbalance is intensified. This makes the production of NUD space seem to be in a situation where it cannot be fundamentally repaired.

4.5 Conclusion

The development of China's NUDs is a typical process of spatial production. NUD is the main spatial organization form for promoting rapid urbanization, which has gradually become an important means of fierce competition between regions. 224 NUDs are located in 157 prefecture-level administrative regions. Among them, 96% are distributed to the east of Hu Line, and 56% are completely within the city's municipal districts. In particular, 64% of NUDs are established within 36 km from their every city center and with an area of 423 km^2 or less. It brings a new geographical imbalance, mainly in two aspects: the production of spatial scales and the production of geographical differences. Various governments dominate the spatial production of NUDs. Geographic space is regarded as the field of power game and rights struggle, and the new political or economic functions formed by administrative division have a dual impact on the spatial production of NUDs. There are also significant scale-level differences in the spatiotemporal evolution of NUDs. The essence of this is the interweaving process of power reconstruction and spatial reorganization, which ultimately leads to new imbalances or inadequate development.

NUD is a comprehensive functional district different from the traditional development zones, and it has a positive effect on promoting regional development and urban–rural integration. First of all, the vitality of small and medium-sized cities based on NUDs should be fully stimulated through administrative divisions and spatial governance in a timely manner, and the spatial structure and functional transformation of NUDs should be improved in the rational prediction of spatial distance and scale. Next, it is necessary to use high-efficient institutional innovation to regulate the power expansion of local governments, and promote an orderly and sustainable development of NUDs on the basis of respecting regional differences. In addition, the state or government departments need to achieve multiple synergies in the governance of NUDs, development zones and municipal districts through power constraints. In the new-type urbanization, NUD should highlight and support the rights to differentiate development. Only when the city is integrated into the system designs according to the fair and just life rights, the standardized and orderly spatial production mode between the practice subjects can be constructed on the basis of "people-oriented" urbanization in the new era.

References

1. Abubakar IR, Doan PL (2017) Building new capital cities in Africa: lessons for new satellite towns in developing countries. Afr Stud 76(4):546–565
2. Brown JA (2019) Territorial (In) Coherence: labour and special economic zones in Laos's border manufacturing. Antipode 51(2):438–457
3. Cartier C (2016) A political economy of rank: the territorial administrative hierarchy and leadership mobility in urban China. J Contemp China 25(100):529–546
4. Che Y (2017) The expansion of cities and the ensuing limitations on the production of space. Sotsiologicheskie Issledovaniya (7):107–115
5. Chen M, Liu W, Lu D (2016) Challenges and the way forward in China's new-type urbanization. Land Use Policy 55:334–339
6. Chen M, Liu W, Tao X (2013) Evolution and assessment on China's urbanization 1960–2010: under-urbanization or over-urbanization? Habitat Int 38:25–33
7. Chen M, Lu D, Zha L (2010) The comprehensive evaluation of China's urbanization and effects on resources and environment. J Geog Sci 20(1):17–30
8. Chen H, Wu Q, Cheng J, Ma Z, Song W (2015) Scaling-up strategy as an appropriate approach for sustainable new town development? Lessons from Wujin, Changzhou, China. Sustainability 7(5):5682–5704
9. Colenutt B, Schaebitz SC, Ward SV (2017) New towns heritage research network. Plan Perspect 32(2):281–283
10. Elden S (2004) Understanding Henri Lefebvre: Theory and the possible. Continuum, London
11. Fang C (2015) Important progress and future direction of studies on China's urban agglomerations. J Geog Sci 25(8):1003–1024
12. Feng K (2016) China's new town and new district development report. Enterprise Management Publishing House, Beijing. [冯奎. 2016. 中国新城新区发展报告. 北京: 企业管理出版社]
13. Firman T (2004) New town development in Jakarta metropolitan region: a perspective of spatial segregation. Habitat Int 28(3):349–368
14. Fu Y, Zhang X (2017) Planning for sustainable cities? A comparative content analysis of the master plans of eco, low-carbon and conventional new towns in China. Habitat Int 63:55–66
15. Fu Y, Zhang X (2018) Two faces of an eco-city? Sustainability transition and territorial rescaling of a new town in Zhuhai. Land Use Policy 78:627–636
16. Gottdiener M (1993) A Marx for our time: Henri Lefebvre and the production of space. Sociol Theory 11(1):129–134
17. Gregory D (1994) Geographical imaginations. Blackwell, Oxford
18. Harvey D (1973) Social justice and the city. Edward Arnold, London
19. Harvey D (1981) The spatial fix: Hegel, Von Thunen, and Marx. Antipode 13(3):1–12
20. Harvey D (1985) The urbanization of capital. Blackwell, Oxford
21. Harvey D (1989) The condition of postmodernity. Blackwell, Oxford
22. Howard E (2007) Garden cities of to-morrow. Routledge, New York
23. Khamaisi R (1998) Building new towns in the formation of a new state of Palestine. Third World Planning Review 20(3):285–308

24. Kuang W, Yang T, Yan F (2018) Examining urban land-cover characteristics and ecological regulation during the construction of Xiong'an New District, Hebei Province, China. J Geog Sci 28(1):109–123
25. Lau JCY, Chiu CCH (2013) Dual-track urbanization and co-location travel behavior of migrant workers in new towns in Guangzhou, China. Cities 30:89–97
26. Lee YS, Shin H (2012) Negotiating the polycentric city-region: developmental state politics of new town development in the Seoul capital region. Urban Stud 49(6):1333–1355
27. Lefebvre H (1991) The production of space (N. Smith, Trans.). Blackwell, Oxford
28. Lefebvre H (2009) State, space, world: Selected essays. University of Minnesota Press, Minneapolis
29. Li L (2015) State rescaling and national new area development in China: the case of Chongqing Liangjiang. Habitat Int 50:80–89
30. Li J, Chiu LHR (2018) Urban investment and development corporations, new town development and China's local state restructuring - the case of Songjiang new town, Shanghai. Urban Geogr 39(5):687–705
31. Liu Y, Chen C, Li Y (2015) Differentiation regularity of urban-rural equalized development at prefecture-level city in China. J Geog Sci 25(9):1075–1088
32. Liu T, Qi Y, Cao G, Liu H (2015) Spatial patterns, driving forces, and urbanization effects of China's internal migration: county-level analysis based on the 2000 and 2010 censuses. J Geog Sci 25(2):236–256
33. Long H (2014) Land consolidation: an indispensable way of spatial restructuring in rural China. J Geog Sci 24(2):211–225
34. Ma LJC (2002) Urban transformation in China, 1949–2000: a review and research agenda. Environ Plan A 34(9):1545–1569
35. Martinez MH (2018) "National Level New Areas" and urban districts: centralization of territorial power relations in China. Chin Polit Sci Rev 3(2):195–210
36. Meir A, Karplus Y (2018) Production of space, intercultural encounters and politics: dynamics of consummate space and spatial intensity among the Israeli Bedouin. Trans Inst Br Geogr 43(3):511–524
37. Merrifield A (2006) Henri Lefebvre: a critical introduction. Routledge, New York
38. Qiu S, Yue W, Zhang H et al (2017) Island ecosystem services value, land-use change, and the National New Area Policy in Zhoushan Archipelago, China. Island Stud J 12(2):177–198
39. Shen J, Wu F (2017) The suburb as a space of capital accumulation: the development of new towns in Shanghai, China. Antipode 49(3):761–780
40. Shields R (1999) Lefebvre, love & struggle: spatial dialectics. Routledge, London
41. Smith N (2010) Uneven development: Nature, capital, and the production of space. University of Georgia Press, Athens
42. Soja EW (1996) Third space: Journeys to Los Angeles and other real-and-imagined places. Blackwell, Oxford
43. Song Y (2005) Influence of new town development on the urban heat island: the case of the Bundang area. J Environ Sci 17(4):641–645
44. Su W, Ye G, Yao S, Yang G (2014) Urban land pattern impacts on floods in a new district of China. Sustainability 6(10):6488–6508
45. Tanabe H (1978) Problems of the new towns in Japan. GeoJournal 2(1):39–46
46. Wang C (2009) Dialectics in multitude: an exploration into/beyond Henri Lefebvre's conceptual triad of production of space. J Geogr Sci (Taiwan) 48(55):1–24. [王志弘. 2009. 多重的辩证: 列斐伏尔空间生产概念三元组演绎与引申. 地理学报(台湾), 48(55): 1–24]
47. Wang Y, Heath T (2010) Towards garden city wonderlands: new town planning in 1950s Taiwan. Plan Perspect 25(2):141–169
48. Wu F, Zhang F, Webster C (2013) Informality and the development and demolition of urban villages in the Chinese peri-urban area. Urban Stud 50(10):1919–1934
49. Xue C, Wang Y, Tsai L (2013) Building new towns in China: a case study of Zhengdong New District. Cities 30:223–232

50. Yang Z, Zhu X, Moodie DR (2015) Optimization of land use in a new urban district. J Urban Plann Dev 141(2):05014010
51. Ye C, Chen M, Chen R, Guo Z (2014) Multi-scalar separations: land use and production of space in Xianlin, a university town in Nanjing, China. Habitat Int 42:264–272
52. Ye C, Chen M, Duan J, Yang D (2017) Uneven development, urbanization and production of space in the middle-scale region based on the case of Jiangsu province, China. Habitat Int 66:106–116
53. Ye C, Ma X, Cai Y, Gao F (2018) The countryside under multiple high-tension lines: a perspective on the rural construction of Heping Village, Shanghai. J Rural Stud 62:53–61
54. Ye C, Zhu J, Li S, Yang S, Chen M (2019) Assessment and analysis of regional economic collaborative development within an urban agglomeration: Yangtze River Delta as a case study. Habitat Int 83:20–29
55. Ye C, Liu Z, Cai W, Chen R, Liu L, Cai Y (2019) Spatial production and governance of urban agglomeration in China 2000–2015: Yangtze River Delta as a case. Sustainability 11(5):1343
56. Zhang T (2000) Land market forces and government's role in sprawl: the case of China. Cities 17(2):123–135
57. Zhang J, Wu F (2008) Mega-event marketing and urban growth coalitions: a case study of Nanjing Olympic new town. Town Plann Rev 79(2/3):209–226
58. Zheng Q, Zeng Y, Deng J, Wang K, Jiang R, Ye Z (2017) "Ghost cities" identification using multi-source remote sensing datasets: a case study in Yangtze River Delta. Appl Geogr 80:112–121
59. Zhuang L, Ye C (2018) Disorder or reorder? The spatial production of State-level new areas in China. Sustainability 10(10):3628
60. Zhuang L, Ye C, Ma W, Zhao B, Hu S (2019) Production of space and developmental logic of New Urban Districts in China. Acta Geogr Sin 74(8):1548–1562. [庄良, 叶超, 马卫, 赵彪, 胡森林. 2019. 中国城镇化进程中新区的空间生产及其演化逻辑. 地理学报, 74(8): 1548–1562]

Chapter 5
Disorder or Reorder? The Spatial Production of State-Level New Areas in China

Abstract With rapid urbanization in the world, new town construction has become prosperous. In particular, new emerging towns in China are unique because of the most significant movement of "building cities". Over four decades of reform and opening-up, this movement has brought about a special development model known as State-level New Area (SLNA) which, like a new town, is causing a growth spurt in national and regional economic development. By applying the critical theory of production of space, this paper gives an overall analysis. SLNAs generate a new expansion pattern of urban space in the regionalization process dominated by governments. To reveal the spatiotemporal evolution logic of SLNA, the framework identifies the main characteristics contributing to spatial production: both bottom-up and top-down project on construction; a sharp and unordered trend of increment in time scale; an unbalanced regional distribution in the sequential order of "Eastern–Western–Northeastern–Central" among regions; complex spatial overlaying with different development zones and administrative divisions; and large-scale spatial expanding. This paper finds that the ongoing growth of SLNAs is a rapid process of spatial production with more contradictions, which is especially marked by tension between disorder and reorder. We hope to provide theoretical reference and practical guidance for the sustainable urbanization and orderly regional development of SLNAs.

5.1 SLNAs and Spatial Production

New towns, whose history dates back at least as far as Ebenezer Howard's Garden cities as the right combination of city and countryside in 1898, are popular around the globe [27]. To solve the urgent need for housing in post-war reconstruction, the UK made the earliest practice of the suburban garden city in 1946 [12, 25], which has had far-reaching impact on Europe, America, Japan, Australia, South Korea, China, and so on [13, 47, 49, 51]. It is a challenge for modern new towns to promote sustainable development of cities [11]. Especially for East Asia, there

---

This chapter is based on [*Sustainability*, Zhuang, L., & Ye, C. (2018). Disorder or reorder? The spatial production of State-Level New Areas in China. *Sustainability*, 10(10), 3628].

© The Author(s), under exclusive license to Springer Nature Singapore Pte Ltd. 2023
C. Ye and L. Zhuang, *Urbanization and Production of Space*, Urban Sustainability, https://doi.org/10.1007/978-981-99-1806-5_5

had been a "mismatch" between western paradigms and local realities because the imported ideas were not properly translated by local planners [54]. The majority of new towns in China have not been built according to planning in an orderly manner and cannot be defined as real cities from a pure feature-and-function perspective [15, 67]. However, to ease the problems of population pressure, traffic congestion, and limited room [14, 26, 28, 66], the new town model is still a natural choice in many areas or countries together with the rapid urbanization. New towns or satellite towns in their true sense should be dynamic, balanced, and self-contained [3, 43], while a current centralized development strategy would be counterproductive [48]. As unreasonable town planning has produced a new objective of "villagers in the city" in the third world, China has also been criticized for many large-scale new town projects [10, 18, 63]. Meanwhile, the rapid development of new towns still brings some challenges to China's new-type urbanization [7], such as social segregation, economic underdevelopment, and environmental degradation [1, 16, 30, 41, 46, 52].

The new town in China, which is often called New Area because of large-scale land use, has special social and institutional context. Compared to other countries, the Chinese State Council and the local governments at various levels play an important role in the process of land finance [5, 9, 29, 35, 36]. This reflects the strong influences of power and capital [23, 40]. Although China achieves the transfer of land use and to realize urban-rural integration, it has encountered lots of trouble in practice [4, 53]. Chinese new town movement contrasts with small-scale urbanization in other developing countries, as well as with suburban sprawl guided by local governments or the private sector in some developed countries. As both a spatial organization and administrative unit, new towns in China must be labeled with different ranks. "Balanced development" is the most outstanding principle of new towns, and the state-level title of SLNA (State-level New Area) represents a profound regional strategy which is closely related to the advantages of "growth poles" and geographic location at national scale [15, 17, 39]. The growth-oriented governments in China still choose the new town as spatial organization to encourage the pursuit of GDP growth. In the 1990s, the transformation from "socialist planned economy" to "market economy" provided local governments with more autonomous decision-making rights and more desire for higher administrative power. However, the explanation for this phenomenon is not enough. In consideration of the conceptual confusion between "new town" (*xin cheng*) and "new area" (*xin qu*), this paper attempts to focus on the special process and dynamic of SLNAs in China.

SLNAs have prominent characteristics of spatial production. Since the reform and opening-up policy, especially since 2000, rapid urban transformation in China has been accompanied with massive new towns. The numbers and sizes of new towns are becoming increasingly larger, reshaping the relationship between city, countryside, and new town, and even university towns have become important methods of spatial production [2, 59]. As new developing spaces, new towns have fueled many uncoordinated urban sprawls via urban land-use change on the outskirts of cities [29, 58]. It is the combination of the land market, local government, urban planners, property developers, and fiscal decentralization that causes rapid urban sprawl in China [20, 64]. The Party-state still plays multiple powerful roles in urban development despite some decentralization of powers from the central to local levels [34, 56]. Facing the problem of regional inequalities, the space of the city-region is a key form of state spatial selectivity [8, 38, 57]. Different from traditional new towns in China, an SLNA, also named National New Area, is a regional comprehensive functional area that is established based on relative administrative divisions and special functional zones, which should be approved by the State Council and undertake important strategic tasks of development and reform [37]. The proliferation of SLNAs in recent years is a typical restructuring regional strategy in rapidly developing China [41]. However, this strategy is not applicable to all kinds of new town [6]. Uneven and insufficient development has become the main feature of China's social contradictions. Therefore, it is an interesting and difficult problem to judge whether the SLNA reflects disorder or reorder of regional development, especially in terms of urban expansion and economic growth. Besides, the existing research is limited in several case studies [33] of local new towns [6, 36, 40, 65]. Little work has been done from a theoretical perspective of spatial production. The paper aims to construct a theoretical framework to reveal the spatiotemporal evolution logic of SLNA and its main characteristics, to provide theoretical reference and practical guidance for sustainable urbanization and orderly regional development.

5.2 Research Methodology

The production of space is a significant critical theory [21, 31, 45] to interpret the development of SLNAs. Building an SLNA is a risky enterprise that requires immense financial investment [19, 67], so it is an interactional process of political, spatial, and economic factors. In other words, the SLNA model mainly promotes spatial urbanization through national political power to realize orderly development between regions, especially economic growth. To some degree, it can be seen as a spatial fix for capital or as a solution to sustain accumulation [44]. Since 1990, the rapid urbanization in SLNAs has witnessed great changes of urban spatial form and structure [24]. Thus, we need a new and strong theory to explain these new areas. The theory of spatial production argues for a trinity of space: ethical appeal is justice, essential attribute is politics, and the major trend is urbanization and globalization [31, 32]. In short, the production of space can be defined as a process that the space is reshaped by capital, power, and class, and then turns into a product of them [60–62]. The establishment of SLNAs is a typical process of spatial production, which embodies the process among different factors. Suburbanization and the decline of central urban areas are an inevitable result of the interaction between capital accumulation and class struggle [21, 22], and uneven development is the key subject of spatial production [45]. However, existing research is limited in several case studies [33], especially those of local new towns [6, 36, 40, 65]. Little work has been done from the theoretical perspective of spatial production. It is necessary to discover the evolving spatiotemporal characteristics and developmental logic of SLNAs based on the theory of spatial production.

In the process of theoretical transformation of spatial production, there are three key methodological principles to use for reference: Firstly, do not forget and give up the starting point. Specifically, do not regard Marx as the "master of thought" and do not be too entangled in details. In accordance with the spirit of dialectics, Marx's doctrine should be regarded as a more complete theoretical system that needs to be critically examined and developed. Secondly, stay critical and free. The methodologist should neither destroy metaphysics nor bind imagination. The task of the methodologist is to identify the tools that can be used to accomplish research and to evaluate them. Therefore, researchers must have a critical spirit. Freedom on methodology is very important and creates the necessary flexibility in the study of geographic phenomena. Thirdly, combine political and economic analysis with literary and artistic techniques. This is embodied in the traditional tool of Marxist political economics and in literature and art, including "image space" such as novels, poems, texts, and pictures. In addition, there is also a combination of the above two methods, typical as David Harvey's *Paris*.

According to the theory of production of space, this paper designs a framework to indicate the developmental logic of SLNA. Figure 5.1 reveals a multidimensional and interactional process of spatial production. T1 phase represents the initial administrative division setting which is an important spatial basis for national comprehensive governance, T2 and T3 reflect the spatial evolution of development zones and new towns, respectively. Development zone is a functional spatial reconstruction based on the administrative division, while new town is based both on the administrative division and development zone. The logic is mainly manifested in four aspects.

5.2.1 The Unique Feature of SLNA

SLNAs are different from traditional new towns (defined as the general name of the relatively independent types of towns newly constructed near cities, also known as satellite towns and new communities) and other development zones, which create new spaces for rapid urbanization. On the one hand, the area of SLNA far exceeds most new towns and development zones. Traditional new towns are mostly less than 100 km^2 in area [49], while SLNAs are all above 460 km^2 and often larger than local administrative divisions at county level. Although there are widespread large-scale development zones, they are almost below their local county-level divisions. On the other hand, SLNAs, different from traditional development zones, are new exclusive spaces with comprehensive functions. Special Economic Zones (SEZs) were established in the late 1970s and early 1980s, and developed in the 1990s, marking further reform and opening-up. During this period, about 18 types of national development zones appeared and brought the total up to 646. The five major types of them are SEZ, Economic and Technological Development Zone (ETDZ), High-Tech Industrial Development Zone (HTIDZ), New Area (NA), and Pilot Free Trade Zone (PFTZ) (Table 5.1). However, SLNA is a unique functional area with complex function including social management and economic development, and the other ones

5.2 Research Methodology

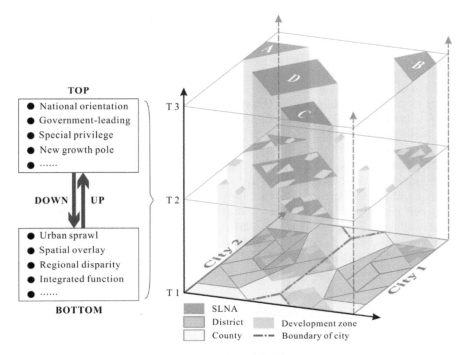

Fig. 5.1 The basic evolution of spatial production of SLNAs

belong to single functional zones which mainly focus on economic or industrial development. Thus, the unique comprehensiveness of SLNAs is the biggest difference from other development zones, showing the significance of regional growth poles in the national space.

Table 5.1 The major types of national developmental zone

Type	Start year	Number	Function	Key words
SEZ	1980	7	Single	Special policy, independent economy
ETDZ	1984	219	Single	Preferential policy, industrial agglomeration
HTIDZ	1988	156	Single	Preferential policy, characteristic Industry
NA	1992	19	Complex	Special policy, regional growth pole
PFTZ	2013	11	Single	Flexible policy on tariff, approval and management

Note Special Economic Zone (SEZ); Economic and Technological Development Zone (ETDZ); High-Tech Industrial Development Zone (HTIDZ); New Area (NA); Pilot Free Trade Zone (PFTZ). Data source: Official Website of China Development Zone, 2018/8/11, www.cadz.org.cn

5.2.2 Spatial Overlay: From Zone to Area and to District

The development of SLNAs is a process of spatial overlay, following the logic from development zone to New Area, and then to administrative district. The word "zone" refers to the space distinguished from adjacent parts by a distinctive feature, "area" refers to the space with indefinite boundary and comprehensive services, "district" refers to the space marked off for administrative or other purposes. To a certain extent, the final solution of spatial overlay is to reconstruct a new government of SLNA and to regain control from the different local governments gradually. Therefore, the developmental process of SLNAs reflects the order of functional evolution: from the single-function "development zone", to the complex-function "New Area", and then to the jurisdictional-function "administrative district". Moreover, there is no mutual repulsion between all zones, areas, and districts. That is, a common space may involve all types of zones, areas and districts. That is why SLNAs can include more different kinds of developmental zones and reshape the process of spatial overlay with administrative division, ETDZ, HITDZ, PFTZ, new town, and so on.

5.2.3 Different Types: Management Model and Spatial Pattern

SLNAs have different management models and spatial patterns. In different stages of development, SLNAs often take corresponding management models in accordance with their own conditions from three major types: Management Committee (*guan wei hui*), Combination of Administrative Region with SLNA (*zheng qu he yi*), and Government (*zheng fu*). The management committee is a dispatched agency to exercise prime powers of development and construction, and social affairs are vested in the local government. In the second model, if the SLNA and administrative division have the same overlapping area, they will build a co-working space for management. As for the last model, with the increasing requirements of urban construction, social management, and service, the SLNA needs to break the administrative division manacles to be integrated into an independent administrative district. Accordingly, the new government approved by the State Council has absolute powers of administrative management in the SLNA. Furthermore, the spatial formation of SLNAs fall into four patterns (Fig. 5.1): "A" means the SLNA is constituted by central districts of one city; "B" refers to a combination of both central districts and peripheral counties from the same city; "C" means the SLNA is constituted by peripheral counties of one city; "D" refers to a combination of both central districts and peripheral counties from different cities. The development order of spatial patterns, A-B-C-D, reflects an urban sprawl by breaking through different borders of administrative divisions. For instance, the shift from "cross-county level" to "cross-prefecture level" has revealed that there is a trend of spatial expansion with long distance and regional cooperation among different cities.

5.2.4 Bidirectional Process: Bottom-Up and Top-Down

The SLNA underlines a mix of bottom-up development appeal (mainly referring to "right" to solve the problems of urban sprawl, spatial overlay, regional disparity, and integrated function) and top-down establishment demand (mainly referring to "power" to realize the purposes of national orientation, government-leading, special privilege, and new growth pole). For one thing, the functional appeals of social management and service are increasing due to the massive establishment of single-function development zones. Local governments need to take more comprehensive measures to integrate the functions of development zones, to change the limitations of various systems and mechanisms. Because of the gradient strategy among regions and the widening regional disparity, local governments must create more new towns as growth poles, to promote regional development. One popular phenomenon is generated: chasing higher administrative titles for more special privileges and supporting policies. A sobering fact about fierce application is that few better new towns or otherwise win a title of "SLNA". For another, SLNA also reflects a political control from top down. Based on the spatial layout of regional strategies, the State Council will select some spaces and corresponding governments to prepare for building an SLNA. After being approved by the State Council over several years, the SLNA should follow a top-down system of administrative control. Not only the strategic orientation of the nation, but also a direct management agency is dispatched by the upper government. For example, the establishment of SLNA must be approved by the State Council of the People's Republic of China (SCPRC) and yield to the unified management of the National Development and Reform Commission of People's Republic of China (NDRCPRC). The SLNA will be subject to the administration of governments at different levels, and level-by-level up-submission. Besides, the most important leadership positions of SLNA are often held by major provincial leaders. In short, the establishment of SLNAs is a typical process of spatial production. Under the transformation appeal of traditional development models, many kinds of single-function development zones need to be transformed into complex-function new areas. Meanwhile, traditional development order has exacerbated regional disparity or disorder, which prompts the state to build a new batch of growth poles to reorder regional development. Therefore, in the bidirectional interaction between "bottom-up" and "top-down", the state will select some exclusive spaces for SLNAs with special privileges and supporting policies that providing local governments with more new spaces for urban expansion.

5.3 The Evolving Spatiotemporal Characteristics of SLNAs

5.3.1 The Temporal Characteristics of SLNAs

SLNA has become a significant development model since the 1990s. In the process of China's rapid spatial expansion by all kinds of development zones, SLNA was first established in the early 1990s and promoted as a new urban area for concentrating on the reform and opening. Since the further policy decision in 1992, the successful model of SEZ has been gradually transferred to SLNA. With the influence of Pudong New Area of Shanghai, China has undergone a new round of building massive new towns across the country. By the end of 2017, China had set up 19 SLNAs, covering a land area of about 22,396 km^2 and a sea area of about 25,800 km^2 (Table 5.2). Among them, just like the first Pudong New Area, the latest Xiong'an New Area approved in 2017 is raising wider international concerns. The reason is not only that the two major SLNAs are important historical strategies made by both of Central Committee of the Communist Party of China (CCCPC) and SCPRC, but also the latter is the only SLNA with main task of "relocating non-capital functions". Therefore, its development planning will exert a strong influence on the construction standards of future SLNAs. To some extent, the four independent municipalities of China have all achieved a strategic layout corresponding to one SLNA.

The establishment of SLNAs has distinguishing stage characteristics. It can be basically divided into three stages (Fig. 5.2): the first one is the Slow Exploration Stage (1990–2009). For 20 years, there were only two SLNAs established in Shanghai and Tianjin with a 16 years gap. At present, the planning area of Binhai New Area is approximately 1.9 times of Pudong New Area. The increased area in this stage accounts for about 16% of the total land area of entire SLNAs, with an annual average of about 205 km^2. The second stage named the Accelerated Development Stage (2010–2013). China had established four new SLNAs for three consecutive years from 2010 to 2013, which kept a land rise of 4249 km^2 and maintained annual growth of 1062 km^2. During the period, the fourth SLNA of Archipelago New Area began to include sea area for the first time. The last stage is the Soaring Explosion Stage (2014–2017). The rest thirteen SLNAs have been continuously set up in only four years. Furthermore, the two same historic peaks of annual increment are intensively reached in 2014 and 2015, which account for more than half of the total SLNAs. It is worth noting that another SLNA, West Coast New Area of Qingdao, has already included a sea area of 5000 km^2 once more.

The establishment of SLNAs is a rapid but unstable process. In terms of the number of SLNAs during the 28-year establishment, there is a consecutive zero growth for 15 years and 89% of the SLNAs were established during the following 8 years. Meanwhile, the average annual increment ratio for the three stages is 1: 2: 6.5. With regard to the planning area of SLNAs, the increased land area accounts for 84.3% of the sum total. Especially, the year of 2014 is a landmark for the establishment of SLNAs. The land expansion scale of new Areas in 2014 reached about 2.8 times that of 2015. In addition, the sea area has officially been included to the overall

5.3 The Evolving Spatiotemporal Characteristics of SLNAs

Table 5.2 Basic information of SLNAs

No.	New area	Approval time	Department	Planning area/km^2	Population/ten thousand	City	Region
1	Pudong	1990.06	CCCPC, SCPRC	1210	547.5	Shanghai	ER
2	Binhai	2006.05	SCPRC	2270	297.0	Tianjin	ER
3	Liangjiang	2010.05	SCPRC	1200	242.6	Chongqing	WR
4	Archipelago	2011.06	SCPRC	1440	97.4	Zhoushan	ER
5	Lanzhou	2012.08	SCPRC	806	16.1	Lanzhou	WR
6	Nansha	2012.09	SCPRC	803	77.8	Guangzhou	ER
7	Xixian	2014.01	SCPRC	882	95.2	Xian, Xianyang	WR
8	Guian	2014.01	SCPRC	1795	77.3	Guiyang, Anshun	WR
9	West Coast	2014.06	SCPRC	2096	180.0	Qingdao	ER
10	Jin Pu	2014.06	SCPRC	2299	158.0	Dalian	NR
11	Tianfu	2014.10	SCPRC	1578	250.3	Chengdu, Meishan	WR
12	Xiangjiang	2015.04	SCPRC	490	134.0	Changsha	CR
13	Jiangbei	2015.06	SCPRC	788	148.2	Nanjing	ER
14	Fuzhou	2015.09	SCPRC	800	155.3	Fuzhou	ER
15	Dian Zhong	2015.09	SCPRC	482	76.1	Kunming	WR
16	Harbin	2015.12	SCPRC	493	36.3	Harbin	NR
17	Changchun	2016.02	SCPRC	499	–	Changchun	NR
18	Ganjiang	2016.10	SCPRC	465	–	Nanchang, Jiujiang	CR
19	Xiongan	2017.04	CCCPC, SCPRC	1770	–	Baoding	ER

Note (1) In addition, the sea areas of Archipelago New Area and West Coast New Area are 20,800 km^2, 5000 km^2, respectively. (2) Central Committee of the Communist Party of China), SCPRC (State Council of the People's Republic of China (CCCPC); Eastern Region (ER); Central Region (CR); Western Region (WR); Northeastern Region (NR). (3) The deadline for population data is 2015

Data source Pudong New Area, www.pudong.gov.cn; Binhai New Area, www.bh.gov.cn; Liangjiang New Area, www.liangjiang.gov.cn; Archipelago New Area, www.zhoushan.cn; Lanzhou New Area, www.lzxq.gov.cn; Nansha New Area, www.gzns.gov.cn; Xixian New Area, www.xixian xinqu.gov.cn; Guian New Area, www.gaxq.gov.cn; West Coast New Area, www.huangdao.gov.cn/n10/index.html; Jin Pu New Area, www.dda.gov.cn; Tianfu New Area, www.cdtf.gov.cn; Xiangjiang New Area, www.hnxjxq.gov.cn; Jiangbei New Area, http://njna.nanjing.gov.cn/; Fuzhou New Area, http://fzxq.fuzhou.gov.cn; Dian Zhong New Area, www.dzxq.gov.cn; Harbin New Area, www.harbin.gov.cn; Changchun New Area, www.ccxq.gov.cn; Ganjiang New Area, www.gjxq.gov.cn; Xiongan New Area, www.xiongan.gov.cn

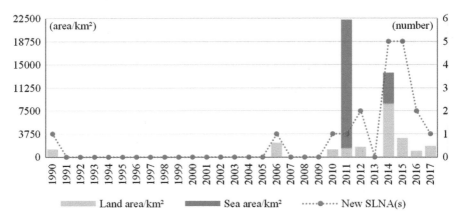

Fig. 5.2 Annual increment about indicators of SLNAs

planning of SLNAs since 2011, which now is about 1.2 times of the land area. Moreover, the establishment of SLNAs is closely related to national development strategy. In particular, China began implementing the state policy of "opening to the outside world" in 1978, and both of the CCCPC and SCPRC put forth the strategy of Pudong development in 1990. As for the year 1992, it is a "watershed" year for China's progressive reform and opening up. The accelerated development zone model has brought about fundamental changes in China since then. With the arrival of 2013, China entered a new era to deepen reform comprehensively, which is regularly followed by explosive establishment of SLNAs.

5.3.2 The Spatial Characteristics of SLNAs

There is a strong order correlation between the spatial layout of SLNAs and the regional division of social-economic development in China. Generally, the great majority of SLNAs are located at the peripheries of regional central cities. While relying on the central old towns, SLNAs are always required to have distinctive functions with the old towns to avoid disadvantages of traditional development paradigms. As a result, SLNA is regarded as a spatial growth pole for regional development and an engine of innovative transformation. The division of the four major regions (Eastern Region, Central Region, Western Region and Northeastern Region), based on the socio-economic development, is a key basis for CCCPC and SCPRC to formulate regional policies (Fig. 5.3a). The earliest two SLNAs are in the Yangtze River Delta Area (YRDA) and the Beijing-Tianjin-Hebei Area (BTHA) of Eastern Region. Soon after that, the approval of the third one marks the SLNA entered the Western Region for the first time. Till then, the earlier 9 SLNAs are all located in the Eastern or Western Region (number is 5 and 4, respectively). Since 2014, the establishment of

5.3 The Evolving Spatiotemporal Characteristics of SLNAs

Jinpu New Area has formally filled the gap of SLNA in Northeastern Region. In the next year, the Xiangjiang New Area indicated that the SLNA began to become a national strategy for Central Region. Since then, the spatial layout of SLNAs has been gradually promoted in the four major economic regions. Currently, the distribution status of SLNAs in the four regions is uneven: 8 for Eastern Region, 6 for Western Region, 3 for Northeastern Region, and 2 for Central Region. Hence, both the development sequence and the quantitative distribution of SLNAs follows the evolving spatial orders of "Eastern Region–Western Region–Northeastern Region–Central Region". At the same time, it also follows a fixed rule of "one province or municipality only corresponds to one SLNA". In addition to the three provinces of Northeastern Region, there are some provinces of the other three major regions that still have no SLNA: 2 of Eastern Region, 5 of Western Region, and 4 of Central Region. Because of the small possibility of establishing SLNA in Beijing, the spatial layout of future new SLNAs will focus much more on the Central and Western regions to promote the inland development and opening-up. However, whether it is possible to break the fixed rule in recent years remains a challenge.

The area size and spatial distance of SLNAs are relatively larger, which is characterized by the rapid expansion of city scale. As far as land area is concerned, every one of all SLNAs is larger than 465 km^2 and more than half of them are intensively below 1000 km^2. Besides, there are six SLNAs with areas ranging from 1000 to 2000 km^2, and the other three super SLNAs are all over 2000 km^2. Surprisingly, marine space on the eastern coast of China has been quickly incorporated into the space of SLNAs, for example, both the fourth and ninth ones have covered a sea area of more than the total land area of all SLNAs. However, in terms of the spatial distance between the management committee of SLNA and the upper administrative center, the vast majority of SLNAs are below 10 km or above 20 km, with 8 in each of the two counterparts, showing a clear polarization trend. The Lanzhou New Area has the longest distance of 54.6 km, and the proportion of SLNAs with distances ranging from 10 to 20 km is only 16 per cent. The area size reflects the spatial potential of local development, while the distance reflects the spatial basis of radiation development. Therefore, the area size and distance are both principal spatial elements for SLNAs. According to the comprehensive data of area and distance, the SLNAs can be classified into three categories (Fig. 5.3b): the circle 1 with smaller area and distance contains more than half of SLNAs, and the circle 2 with larger area and distance contains a medium proportion, and the circle 3 with smaller area but larger distance contains the least. Consequently, SLNAs with large areas but different distances tend to be closer to the regional central cities, such as municipalities, capitals of provinces, and large and medium-sized cities, for large-scale urban expansion. Since the integration of new town and transportation is a crucial step in building a sustainable city [55], most of these selected cities belong to the node cities of the main traffic arteries in regions. With the changing scale and distance, the combination of SLNA is becoming increasingly diversified, which is full of contradictions between internal inadequate development and external imbalanced development. This brings a huge challenge for social management and spatial governance. At the same time, SLNA

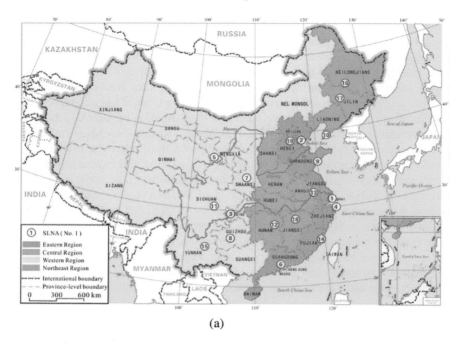

(a)

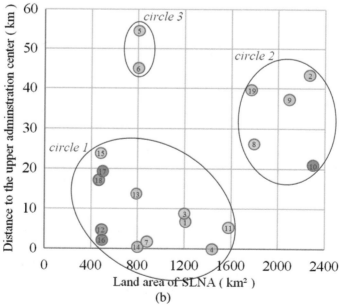

(b)

Fig. 5.3 The spatial distribution and indicators of SLNAs in China. *Note* Numbers from 1 to 19 in this figure are in a one-to-one correspondence with Table 5.2

has gradually shown a more salient trend of "closer distance" and "larger scale", and it tends to be distributed along the coast or river.

China is facing growing management risk from SLNAs due to their establishment pattern which has complexity of administrative divisions and diversity of development zones. There are many management challenges of "spatial overlay" both inside and outside of the SLNAs. On the one hand, most SLNAs are not administrative entities with administrative management functions. However, their orientation of comprehensive functions also requires multiple executive powers. Therefore, the state should appropriately promote the integration of development zones and SLNAs by means of administrative division adjustment. At the same time, it is necessary to achieve multidimensional coordinated management on different spaces through institutional innovation. Otherwise, there will be much management conflicts and coordination difficulties between the SLNAs and their intersections with county-level divisions (including the districts under the jurisdiction of cities, the cities at county level, and the counties). It is about more than 3 county-level divisions in every SLNA, while Tianfu New Area and Xixian New Area have each 7 ones. In addition, the coordination difficulties also come from the upper administrative division. Especially since 2014, the "spatial overlay" between SLNAs and Administrative divisions has upgraded from "cross-county-level divisions" to "cross-prefecture-level divisions", which means that the management coordination is involved in different prefecture-level divisions and their subordinate ones. On the other hand, every SLNA includes different levels, types, or numbers of single-function development zones and even some ones with orientations at state level. Many functional zones basically adopt the management model with dispatched committees, so each of the different zones within SLNAs is often overlaid with other ones. In summary, there is a prominent "multi-level, multi-type and multi-quantity" characteristic of "spatial overlay" between SLNAs, administrative divisions, and functional zones.

5.4 Discussions: The Tension Between Disorder and Reorder

The government-led mode of SLNAs is bound up with the evolutionary orders of regional strategies in China. There is a logical consistency between the establishment of SLNAs and the strategic consequence of four major regions (Fig. 5.4). In the first place, it is "from the south to the north" of Eastern Region. After the reform and opening-up, the establishment of SEZ provided the eastern coast of China with a rapid development. The economy of Eastern Region even through the country has gradually assumed the spatial layout of "south faster and north slower". As national economic center, Shanghai's economic contribution rate in domestic growth was declined year by year from 7.8% in 1978 to 4.19% in 1990. Likewise, another economic center of Tianjin just contributed 17.88% to the GDP of BTHA. Consequently, the two major SLNAs established before 2010 show strategic significance for promoting the economic development of east coast from south to north. Secondly, "from east to west" means a turn from the Eastern Region to the Western Region. After entering the twenty-first century, the increasing regional disparities between Eastern Region and the other regions have received increased attention. With the development of regional strategies (including the China's Western Development, the Revitalization of Northeast China, the Rising of Central China, and the Advancement of Eastern Region), the Central, Western and Northeastern regions have achieved a rapid progress, while the growth rate of Eastern Region was lower than the other three ones for the first time in 2008. So, the subsequent establishment of SLNAs in the Western Region is aiming to reduce the huge regional disparities between regions. Apart from the above two aspects, "from east–west to nation" is a shift from the east–west direction to the whole country. Since 2014, China's economy has entered a stage of "new normal" (*xin chang tai*) with low growth rate, thus cultivating new growth poles for regional development is to be a key reason for soaring SLNAs. Another growing pressure comes from the "cliff-breaking" decline of economy in the Northeastern Region, leading to the first phenomenon of negative economic growth, only negative 9 per cent of average GDP growth rate in 2016, among regions since the reform and opening-up. As a result, the Northeastern Region set up SLNAs in preference to the Central Region. At present, each of three provinces of Northeastern Region has one SLNA, and the Central Region is also advancing.

The establishment of SLNAs has not well achieved orderly development among regions. As exclusive spaces, SLNAs has carried the vital mission of national spatial and regional strategies for more than 20 years. The establishment of SLNAs has always been accompanied by the purpose of balanced development, which is more often expressed as the readjustment from "disorder" to "reorder". However, whether it is "south-north direction" or "east–west direction", their unbalanced development

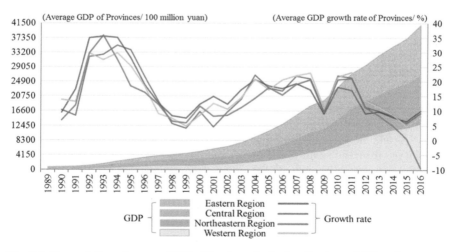

Fig. 5.4 The evolution of average GDP of Provinces in the four Regions of China. *Data source* China Statistical Yearbook [36]

has not break through the existing difficulties. On the contrary, the economic gap between different regions continues to widen, such as the economic downturn in the Northeastern Region in recent years. At the same time, the development level of different SLNAs also differs greatly, and the GDP of some SLNAs does not match well with their corresponding area scale. For example, Guian New Area, jointly built by two prefecture-level cities of Guiyang and Anshun, has a planned area of 1795 km^2 but with GDP about 17,060 million yuan in 2014, contributing only 1.6% to Guizhou Province. Another example is that the Lanzhou New Area, with the longest spatial distance, ranked first-lowest with 12,550 million yuan in 2014. It is also crowned with the title of "Ghost City" because of the high proportion of housing vacancy and farmland losses. In the same year, the existing SLNAs mostly contributed less than 10% to the upper provincial GDP. The Tianfu New Area and Dianzhong New Area are 7.3% and 7.5%, respectively, which are all below the average level of their province. Due to the later establishment of other SLNAs and the municipality system, the Pudong New Area and Binhai New Area contributed as high as 31.6% and 56.1%, respectively [58]. As the two oldest established SLNAs, both have achieved rapid growth in more than 20 years (Fig. 5.5). Since 2008, the GDP of Binhai New Area has surpassed the Pudong New Area for the first time and has been the leader of all SLNAs until now, and the annual economic contribution rate has continuously maintained over 50 per cent. That is why Binhai New Area has been seen as the third growth pole of China following the Shenzhen SEZ and Pudong New Area. It was also labeled as the first SLNA which hit GDP above 1 trillion yuan in 2016, reported by the Xinhua News Agency. However, recently, some China's northern provinces, such as Liaoning and Nei Mongol, cut their GDP growth and admitted fudging key economic numbers. More importantly, the northern Binhai New Area of Tianjin, as the best SLNA, has substantially reduced its 2016 GDP growth data by 33% (24% below Pudong New Area at the same year) to 665.4 billion yuan from the initial official figure of 1000.2 billion yuan. After correcting the actual GDP, it restored the original disorder between the north and the south in China. Before this, the "8·12" catastrophic fire and explosion accident, occurred in the Binhai New Area of Tianjin in 2015, causing serious casualties, which led to a heated discussion over the hidden danger and governance crisis in the development of SLNAs. In a word, we should rethink deeply about the development model of SLNAs. For example, the state should make necessary regulations on the city planning rights of local governments. Besides, the major cities should promote the orderly and sustainable development of SLNAs based on respecting regional differences. Of course, SLNAs must accelerate the transformation from inclined economic functions to complex functions.

5.5 Conclusions

Because of more significant government-led intervention, the SLNA model in China is greatly different from new towns of other countries regarding planning, construction, and management, and is different from new towns itself. The spread campaign

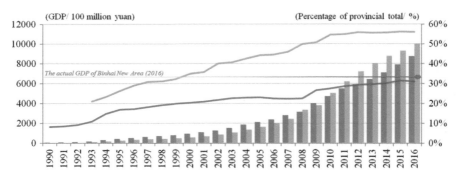

Fig. 5.5 Comparison of GDP between the best two SLNAs. *Data source* Statistical Yearbook of Shanghai Pudong New Area [42], Statistical Yearbook of Tianjin Binhai New Area [50]

of new town establishment aims to relieve the pressures of city center by the polycentric planning and spatial expansion. However, the SLNA has its own peculiarity. SLNAs with large-scale planning are seen as important spatial carriers for regional development that should be explored in an orderly fashion. In fact, most SLNAs are not spontaneous productions but a significant mismatch between planning and practice influenced by state powers, resulting in a disorderly urban sprawl to some extent. Compared to the traditional new town, an SLNA should be understood as a larger city-level administrative district rather than a town. Compared to the traditional development zone, SLNA is a larger special area with complex function that should be responsible for integrating the single-function development zones inside.

The development of SLNAs is a rapid process for the production of space in China. In recent years, the surge of SLNAs has been a major strategic support for the practice of a new type of urbanization. This implies that spatial production will continue, and it seems multidimensional and interactional as it is a product of the spatiotemporal characteristics of SLNAs as a whole. On the temporal scale, the establishment of SLNAs is unstable, which is experienced in three stages: slow exploration, accelerated development, and soaring explosion. On the spatial scale, the spatial layout of SLNAs has a strong order correlation with the four major regions of China, of which there is an evolutionary order of "Eastern–Western–Northeastern–Central". It has an entanglement between disorder and reorder of regional development. Meanwhile, the area of one SLNA is generally larger than a traditional new town. These new areas are more likely getting close to regional hub cities for large-scale spatial production, even expanding to ocean space. Based on the complicated and varied development zones and administrative divisions, the comprehensive areas of SLNAs are making a spatial overlay at many different spatiotemporal levels and scales. However, SLNAs represent a more intricate form of spatial production with function from single to complex and development from local to regional.

Rapid urbanization and economic growth have generated a tension between disorder and reorder in the recent decades. The SLNA is mostly established in a tough adjustment period of regional development. There is no doubt that the building of

SLNAs will certainly stimulate rapid growth in the urbanization and economy in the short term. However, the existing unbalanced development between regions can be embodied in the development sequence and the quantitative distribution of SLNAs. The different SLNAs from different regions have different levels of development. First, some SLNAs are less-effective drivers for regional development. Second, many local governments are still focusing more on economic growth rather than institutional innovation of SLNAs. Third, skepticism surrounding SLNAs is revealing a disorder of unbalanced and inadequate development, for instance in the Binhai New Area and Lanzhou New Area. Of course, SLNAs are the top spaces for new expansion and exploration at the state level, because they are both special areas of comprehensive functions and pilot areas of reform. Although it is certainly positive to present the SLNAs as a panacea for transformation of development zones, the realization is less so. Essentially, such tension also represents a mismatch between spatial production and spatial fix. Understanding the SLNAs from the perspective of spatial production is useful if we understand the tension between disorder and reorder of regional development in China. If the logic of spatial production mainly dominated by government in the process of urbanization cannot be changed, the tension and disorder will continue and even deteriorate, even if a new spatial pattern is produced.

References

1. Abubakar IR, Doan PL (2017) Building new capital cities in Africa: lessons for new satellite towns in developing countries. Afr Stud 76(4):546–565
2. Addie JPD, Keil R, Olds K (2015) Beyond town and gown: universities, territoriality and the mobilization of new urban structures in Canada. Territory Polit Governance 3(1):27–50
3. Atash F, Beheshtiha YSS (1998) New towns and their practical challenges: the experience of Poulad Shahr in Iran. Habitat Int 22(1):1–13
4. Chen M, Ye C (2014) Differences in pattern and driving forces between urban and rural settlements in the coastal region of Ningbo, China. Sustainability 6(4):1848–1867
5. Chen X, Wang L, Kundu R (2009) Localizing the production of global cities: a comparison of new town developments around Shanghai and Kolkata. City Community 8(4):433–465
6. Chen H, Wu Q, Cheng J, Ma Z, Song W (2015) Scaling-up strategy as an appropriate approach for sustainable new town development? Lessons from Wujin, Changzhou, China. Sustainability 7(5):5682–5704
7. Chen M, Liu W, Lu D (2016) Challenges and the way forward in China's new-type urbanization. Land Use Policy 55:334–339
8. Chen M, Gong Y, Li Y, Lu D, Zhang H (2016) Population distribution and urbanization on both sides of the Hu Huanyong Line: answering the Premier's question. J Geog Sci 26(11):1593–1610
9. Chen M, Liu W, Lu D, Chen H, Ye C (2018) Progress of China's new-type urbanization construction since 2014: a preliminary assessment. Cities 78:180–193
10. Cho SE, Kim S (2017) Measuring urban diversity of Songjiang New Town: a re-configuration of a Chinese suburb. Habitat Int 66:32–41
11. Colenutt B, Schaebitz SC, Ward SV (2017) New towns heritage research network. Plan Perspect 32(2):281–283
12. Daniels TL (2016) Practicing utopia: An intellectual history of the new town movement. Plan Perspect 31(4):674–675

13. Desponds D, Auclair E (2016) The new towns around Paris 40 years later: new dynamic centralities or suburbs facing risk of marginalization? Urban Stud 54(4):862–877
14. El-Shakhs S (1994) Sadat city, Egypt and the role of new town planning in the developing world. J Architectural Plann Res 11(3):239–259
15. Feng K (2016) China's new town and new district development report. Enterprise Management Publishing House, Beijing. [冯奎. 2016. 中国新城新区发展报告. 北京: 企业管理出版社]
16. Firman T (2004) New town development in Jakarta metropolitan region: a perspective of spatial segregation. Habitat Int 28(3):349–368
17. Fu Y, Zhang X (2017) Planning for sustainable cities? A comparative content analysis of the master plans of eco, low-carbon and conventional new towns in China. Habitat Int 63:55–66
18. Glover WJ (2012) The troubled passage from 'village communities' to planned new town developments in mid-twentieth-century south Asia. Urban Hist 39:108–127
19. Golany G (1976) New-town planning: principles and practice. Wiley, New York
20. Gu C, Zeng F, Qiu Y, Ye S (1999) Growth of new designated cities in China. Chin Geog Sci 9(2):97–106
21. Harvey D (1973) Social justice and the city. Edward Arnold, London
22. Harvey D (1985) The urbanization of capital. Blackwell, Oxford
23. He S, Lin GCS (2015) Producing and consuming China's new urban space: state, market and society. Urban Stud 52(15):2757–2773
24. He S, Qian J (2017) From an emerging market to a multifaceted urban society: urban China studies. Urban Stud 54(4):827–846
25. Heraud BJ (1966) The new towns and London's housing problem. Urban Stud 3(1):8–21
26. Heraud BJ (1968) Social class and the new towns. Urban Stud 5(1):33–58
27. Howard E (2007) Garden cities of to-morrow. Routledge, New York
28. Khamaisi R (1998) Building new towns in the formation of a new state of Palestine. Third World Plan Rev 20(3):285–308
29. Lee YS, Shin H (2012) Negotiating the polycentric city-region: developmental state politics of new town development in the Seoul capital region. Urban Stud 49(6):1333–1355
30. Lau JCY, Chiu CCH (2013) Dual-track urbanization and co-location travel behavior of migrant workers in new towns in Guangzhou, China. Cities 30:89–97
31. Lefebvre H (1991) The production of space (N. Smith, Trans.). Blackwell, Oxford
32. Lefebvre H (2009) State, space, world: selected essays. University of Minnesota Press, Minneapolis
33. Li L (2015) State rescaling and national new area development in China: the case of Chongqing Liangjiang. Habitat Int 50:80–89
34. Ma LJC (2002) Urban transformation in China, 1949–2000: a review and research agenda. Environ Plan A 34(9):1545–1569
35. Nasution AD, Harisdani DD, Napitupulu PP (2017) The implementation of aerotropolis concept on new town planning and design in Mebidangro, Sumatera Utara. In: 1st annual applied science and engineering conference (AASEC), vol 180, p 012293
36. National Bureau of Statistics of China (2017) China Statistical Yearbook. China Statistics Press, Beijing. [中国国家统计局. 2017. 中国统计年鉴. 北京: 中国统计出版社]
37. National Development and Reform Commission of People's Republic of China (2016) China's State-level New Area development report. China Planning Press, Beijing. [国家发展和改革委员会. 2016. 国家级新区发展报告. 北京: 中国计划出版社]
38. Qin C, Ye X, Liu Y (2017) Spatial club convergence of regional economic growth in inland China. Sustainability 9(7):1189
39. Ricks B (1970) New town development and the theory of location. Land Econ 46(1):5–11
40. Schroeder P (2014) Assessing effectiveness of governance approaches for sustainable consumption and production in China. J Clean Prod 63:64–73
41. Schroeder PM, Chapman RB (2014) Renewable energy leapfrogging in China's urban development? Current status and outlook. Sustain Cities Soc 11:31–39
42. Shanghai Pudong New Area Statistical Bureau (2016) Statistical yearbook of Shanghai Pudong New Area. China Statistic Press, Beijing. [上海市浦东新区统计局. 2016. 上海浦东新区统计年鉴. 北京: 中国统计出版社]

43. Shaw A (1995) Satellite town development in Asia: the case of new Bombay, India. Urban Geogr 16(3):254–271
44. Shen J, Wu F (2017) The suburb as a space of capital accumulation: The development of new towns in Shanghai, China. Antipode 49(3):761–780
45. Smith N (2010) Uneven development: nature, capital, and the production of space. University of Georgia Press, Athens
46. Song Y (2005) Influence of new town development on the urban heat island: the case of the Bundang Area. J Environ Sci 17(4):641–645
47. Steinicke E, Čede P, Löffler R (2012) In-migration as a new process in demographic problem areas of the Alps. ghost towns vs. Amenity settlements in the Alpine border area between Italy and Slovenia. Erdkunde 66(4):329–344
48. Sui D (1995) Spatial economic impacts of new town development in Hong Kong: a GIS-based shift-share analysis. Socioecon Plann Sci 29(3):227–243
49. Tanabe H (1978) Problems of the new towns in Japan. GeoJournal 2(1):39–46
50. Tianjin Binhai New Area Statistical Bureau (2016) Statistical yearbook of Tianjin Binhai New Area. China Statistic Press, Beijing [天津市滨海新区统计局. 2016. 天津滨海新区统计年鉴. 北京: 中国统计出版社]
51. Tuppen JN (1983) The development of French new towns: an assessment of progress. Urban Stud 20(1):11–30
52. Tzoulas K, James P (2010) Peoples' use of, and concerns about, green space networks: a case study of Birchwood, Warrington new town, UK. Urban Forestry Urban Greening 9(2):121–128
53. Wang A, Chan EHW, Yeung SCW, Han J (2017) Urban fringe land use transitions in Hong Kong: from new towns to new development areas. Urban Transitions Conf 198:707–719
54. Wang Y, Heath T (2010) Towards garden city wonderlands: new town planning in 1950s Taiwan. Plan Perspect 25(2):141–169
55. Wei H, Mogharabi A (2013) Key issues in integrating new town development into urban transportation planning. Procedia Soc Behav Sci 96:2846–2857
56. Wu F (2016) China's emergent city-region governance: a new form of state spatial selectivity through state-orchestrated rescaling. Int J Urban Reg Res 40(6):1134–1151
57. Wu F (2016) State dominance in urban redevelopment: beyond gentrification in urban China. Urban Aff Rev 52(5):631–658
58. Xue C, Wang Y, Tsai L (2013) Building new towns in China: a case study of Zhengdong New District. Cities 30:223–232
59. Ye C, Chen M, Chen R, Guo Z (2014) Multi-scalar separations: land use and production of space in Xianlin, a university town in Nanjing, China. Habitat Int 42:264–272
60. Ye C, Chen M, Duan J, Yang D (2017) Uneven development, urbanization and production of space in the middle-scale region based on the case of Jiangsu Province, China. Habitat Int 66:106–116
61. Ye X, He C (2016) The new data landscape for regional and urban analysis. GeoJournal 81:811–815
62. Ye X, Liu X (2018) Integrating social networks and spatial analyses of the built environment. Environ Plan B 45(3):395–399
63. Zacharias J (2005) Generating urban lifestyle: the case of Hong Kong new-town design and local travel behaviour. J Urban Des 10(3):371–386
64. Zhang T (2000) Land market forces and government's role in sprawl: the case of China. Cities 17(2):123–135
65. Zhang J, Wu F (2008) Mega-event marketing and urban growth coalitions: a case study of Nanjing Olympic New Town. Town Plann Rev 79(2/3):209–226
66. Zhang H, Zheng W (2007) Geographical agglomeration of Chinese manufacturing industries. Chin J Population Resour Environ 5(2):3–11
67. Ziari K (2006) The planning and functioning of new towns in Iran. Cities 23(6):412–422

Chapter 6
Spatial Production of National High-Tech Industrial Development Zones in China

Abstract Science parks are popular in most countries of the world. In China they have taken the form of National High-Tech Industrial Development Zones (NHTIDZs), which have demonstrated special spatiotemporal characteristics over the past thirty years. NHTIDZs, as exclusive spaces, epitomize the close relationship between governmental power and urbanization, and have become an organizational form of the production of space. However, little research has been carried out into the spatial production of China's NHTIDZs. Based on the theory of production of space, this article designs a framework for identifying the interactions between governmental power, NHTIDZs, and urbanization. We find there are two main characteristics of the changing imbalance between time and space: a rapid and unstable centralizing trend and an extremely uneven spatial distribution. The NHTIDZ, as a spatial organization pattern of urbanization, is dominated by governmental power. Because of rapid spatial expansion and great policy privileges, national-level NHTIDZs have become targets for governments at all levels. The purpose of the central government is to promote urbanization by expanding high-technology zones nationwide; therefore, urbanization in China is becoming a process of spatial production.

6.1 NHTIDZs and Spatial Production

With the evolution of globalization and the spread of planetary urbanization, urban studies are becoming multi-scalar and cross-disciplinary [6, 7, 9, 69]. Urbanization in China has evolved as a specifically modernized endogenous logic [23, 68]. The triple contemporary processes of decentralization, globalization and marketization have dramatically influenced the national land use and land cover changes, fostering urban sprawl at the urban fringe areas [59]. And this may also be the greatest human-resettlement experiment in history [2]. Since its 1978 Reform and Opening-up Policy, China's transformational urbanization and economic growth have been gaining increasing attention throughout the world. The accelerated out-migration of

---

This chapter is based on [*Land Use Policy*, Zhuang, L., & Ye, C. (2020). Changing imbalance: spatial production of National High-Tech Industrial Development Zones in China (1988–2018). *Land Use Policy*, 94, 104512].

© The Author(s), under exclusive license to Springer Nature Singapore Pte Ltd. 2023
C. Ye and L. Zhuang, *Urbanization and Production of Space*, Urban Sustainability,
https://doi.org/10.1007/978-981-99-1806-5_6

rural labors under urban–rural dual-track system has brought huge obstacles to the improvement of land use efficiency in China [35]. Most literature on China's urbanization has focused on the government's role, the rural–urban relationship, urban transformation, etc. [28, 36, 37, 43, 44, 53]. Land–industry development is a key solution for the housing–industry disjuncture which important causative factor is the low degree of population–industry coordination [17]. At the same time, high-tech development around the world, especially in some emerging science parks (as one kind of spatial and industrial organization), have gradually become an important policy tool to promote local and regional development [16, 31, 56]. More importantly, the proliferation of High-Tech Industrial Development Zones (HTIDZs) in China is very closely related to its rapid urbanization [20, 52, 58]. We need to examine the impact of changes in human socio-economic activities on land-use changes and related policies from the perspective of NHTIDZs [38, 71].

National High-Tech Industrial Development Zones (NHTIDZs), rather than regional HTIDZs, are playing an increasingly crucial role in the production and exploitation of urban space by governments at all levels. The concept of HTIDZs can be traced back to at least the Stanford Industrial Park in the United States, a predecessor of Silicon Valley, established in the 1950s. The Science Park is significant in its encouragement of the creation and transformation of technology through the production of knowledge-based contiguous spaces. Driven by this successful exemplar, many countries such as the UK, Greece, Japan, India and China embarked on a long period of constructing science parks [3, 14, 33, 45, 57]. NHTIDZs in China are often regarded as exclusive spaces with policy privileges and management advantages granted by national or local governments, and located on the periphery of urban spaces that have a complex relation with technologies, universities, research institutions, and governments [19, 21, 51].

The strategic coupling of high technology industries in global production networks is a significant reflection of the changing scales between globalization and localization [63, 64]. Many researchers who work on HTIDZs argue that the proximity of geographical location to high-tech talent is helpful for knowledge exchange [4, 18, 29, 30], but also has a negative impact on innovation because of the problem of lock-in [10]. Furthermore, policymaking is important in developing NHTIDZs [1, 11, 22, 32]. Case studies are limited to several NHTIDZs at the city-level, such as Zhangjiang of Shanghai and Zhongguancun of Beijing [15, 58, 66, 67]. Little work has been done to summarize and explain the spatiotemporal process of NHTIDZs in the past thirty years, especially that focusing on urbanization from the perspective of the production of space [70].

Urbanization in China is experiencing a great transformation from physical space to social-economical space, which is a diversified process of spatial production. The current land use problems in China are basically produced by the fast urban–rural transformation [39]. At the same time, many spatial organizations such as the Economic and Technical Development Zone, University Town, New Socialist Countryside, and New Area/District have emerged during this rapid process of urbanization [54, 60, 61, 71]. Of the various types of development zones, HTIDZs have played a key supporting role in the process of urbanization. Although it is more difficult than ever in the globalizing era to govern markets by direct policy interventions [62], the urbanization pattern and NHTIDZs in China is dominated by government or political power. The redevelopment of China's urban industrial land in the forefront of neo-liberalization is full of inconsistencies, involving not only conflicts between neoliberal practices and social resistances, but also tensions between central and local governments [24]. This perspective, looking at the production of space to explain the interaction between urbanization and NHTIDZs, is interesting and important, especially in connection with the explosive doubled-growth of NHTIDZs after 2010. We focus on solving the following questions: what theoretical logic does China's NHTIDZs have in the government-led urbanization, and what role does it play in optimizing the allocation of urban–rural land resources? Therefore, our innovation and contribution are: this article presents a framework to identify the interactions between governmental power, NHTIDZs, and urbanization, and to explain the spatiotemporal process and characteristics of NHTIDZs involved in China. Furthermore, it puts forward policy suggestions on how to realize the intensive and economical use of land spaces in the historical logic of urban–rural development.

6.2 A Framework Explaining the Interaction Between Government, NHTIDZs and Urbanization

China's urbanization is mainly concerned with the intricate interactions between time–space and power. For several decades, China has been on a course to catch up with the advanced countries in urbanization through the rapid spatial expansion of various existing development zones. The "space" shows a salient exclusion of, and encroachment on, "time" by means of governmental power. It is therefore an urgent task to interpret the logic between government, NHTIDZs, and urbanization. As illustrated in the visualization of framework in Fig. 6.1, the development of NHTIDZs in China is a process of systematic interactions between three main factors: government, NHTIDZs, and urbanization, which correspond to the macro-scale, micro-scale, and city-scale respectively. It should be noted that the framework highlights a bi-directional interaction between adjoining factors.

The three main factors have important impacts on the interactive links between different scales. On the macro-scale, the government is characterized by a hierarchical top-down, nationwide power constitution. In general, the central government, guided by national opinions, always determines the levels of HTIDZs. It can select some special spaces to create new NHTIDZs or upgrade lower-level existing ones. The government, with its dominant power, is also endeavoring to promote urbanization by the production of space. On the micro-scale, the NHTIDZs, as sub-departments of government, are designated and managed by the government and, in this sense, are an extension of bureaucratization. This contributes to the centralization of governmental power and the circulation of policy resources in order to realize the disposition of authority or the shift of power in a national space. As a spatial organization

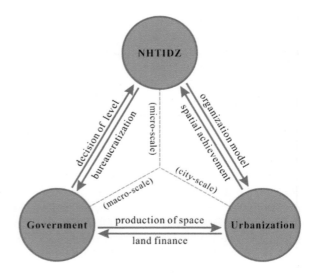

Fig. 6.1 The framework about interaction between government, NHTIDZ and urbanization

model, the NHTIDZs continue to turn into urban space. Given the development advantages of policy, location, and land controlled by the central government, when it comes to property development, the NHTIDZs also form a hierarchical management system and operating mechanism to build a foundation for the spatial production of urbanization. The production of this organization model often follows a logical process of space evolution: from "development zone" thru "built-up area" to "urban area." On the city-scale, urbanization in China is represented by large-scale urban sprawl—a land-centered process driven by land finance [65], which provides the government with financial support. Such government-dominated spatial production can help cities realize the spatial achievement of urban development and construction in NHTIDZs.

Following substantial improvements in infrastructure, economy, population, employment, etc., there would be a rapid development of NHTIDZs that presents a picture of an accelerating urbanization rate as well as the agglomeration of spatial elements. As a result, the larger expansion of urban space generates a stronger impetus for governmental decisions, which in turn prompts a new round of progressive spatial production spurred by the state. In this case not only would space be producing social relationships, but social relationships would also be reshaping space, which is an entity of the social process and spatial pattern [26]. Simultaneously, a variety of contradictions, such as the wealth gap, individual rights, and the human-earth relationship, is produced in the production of space.

6.3 The Evolving Relations Between NHTIDZs and Urbanization

6.3.1 The Temporal Characteristics of NHTIDZs

With their state-led direction and government-oriented management, China's NHTIDZs are remarkable in that they were established on a centralized schedule. We have divided the process of the NHTIDZs' development into four sequential stages (Fig. 6.2), comprising the Experimental Exploration Stage (1988–1990), the Rapid Startup Stage (1991–1992), the Consistent Stagnation Stage (1993–2008), and the Accelerated Growth Stage (2009–). The centralized establishment is reflected primarily in staged increments and annual increments. On the one hand, there have been two rounds of rapid centralization in the development of the NHTIDZs, corresponding to the Rapid Startup Stage and Accelerated Growth Stage. Their impressive increment values at each stage are 51 and 114 respectively, which account for 98% of the total 168. The second round of rapid centralization is proving stronger and more enduring than the first round. On the other hand, while the process has at times fluctuated, more than four-fifths of the NHTIDZs were intensively established in the five years of 1991, 1992, 2010, 2012, and 2015. The three historic peaks of annual increment are relevant to the opening year of the eighth Five-year Plan, as well as to each of the last two years of the Eleventh and Twelfth Five-year Plans. However, the annual increments declined, or stopped, altogether for two new NHTIDZs during the 16 years from 1993 to 2008. This shows that the establishment of NHTIDZs follows a rapid but unstable process. Since then, the changing imbalance of NHTIDZs at the country level has remained consistent with the provincial (Fig. 6.3) and regional levels, especially in terms of average value. Jiangsu, Guangdong, and Shandong are the three provinces with the greatest number of NHTIDZs, as they were all established between 1991–1992 and 2009–2018. As far as regions are concerned, the annual NHTIDZ increment trend in the Yangtze River Delta Area (YRDA) (including the provincial-level administrative districts of Shanghai, Jiangsu, Zhejiang, and Anhui) fits more closely with the country level. In other words, the number of NHTIDZs built over time reflects the typical characteristic of centralized establishment at different levels.

6.3.2 The Spatial Characteristics of NHTIDZs

Due to difference in geographical space or location in China, the spatial distribution of NHTIDZs has two important characteristics of spatial agglomeration and spatial disparity that are also correlated with the hierarchical arrangement of cities (Fig. 6.4). The number of NHTIDZs has increased significantly since 1991. The first was built in the capital Beijing, the center of political power in China, during the period of Experimental Exploration. Subsequently, the first round of large-scale

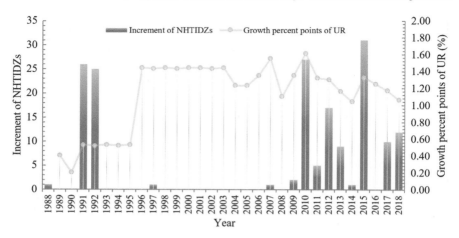

Fig. 6.2 The evolution of NHTIDZs and the growth rate of Urbanization Rate (UR) in China. *Data source* [47, 48]

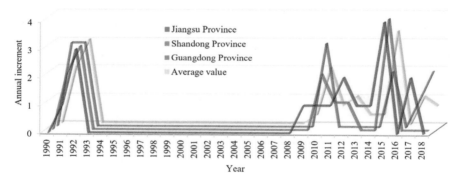

Fig. 6.3 The annual increment of NHTIDZs built in the main provinces. *Data source* [48]

NHTIDZ construction was mainly distributed in the country's provincial capitals and large and mid-sized cities in 1991 and 1992. There then followed a preliminary trend of spatial clustering in the YRDA and Pearl River Delta Area (PRDA), where the foundation for the subsequent spatial pattern of NHTIDZs was built (see the map for 1992 in the upper left of Fig. 6.4). From 1993 to 2010, 31 new NHTIDZs were built in 20 provinces, with several emerging cluster regions such as the Beijing-Tianjin-Hebei Region (BTHR) and Shanxi Province, while the former two largest clustering areas are still growing (see the map for 2010 in the upper right of Fig. 6.4). There has not been any long-term change after this point, except that the first agricultural NHTIDZ and a new technical NHTIDZ in Ningbo City were founded in 1997 and 2007 respectively. In the early 2010s, the new batch of NHTIDZs was focused for the most part on expanding the existing agglomeration areas rather than forming new ones (see the map for 2014 in the lower left of Fig. 6.4). Nevertheless,

33. Li L, Hu P, Zhang L (2004) Roles, models and development trends of hi-tech industrial development zones in China. Int J Technol Manage 28(3–6):633–645
34. Li Y, Wu W, Liu Y (2018) Land consolidation for rural sustainability in China: practical reflections and policy implications. Land Use Policy 74:137–141
35. Li Y, Liu Y, Long H, Cui W (2014) Community-based rural residential land consolidation and allocation can help to revitalize hollowed villages in traditional agricultural areas of China: evidence from Dancheng County, Henan Province. Land Use Policy 39:188–198
36. Lin GCS (1999) State policy and spatial restructuring in post-reform China, 1978–95. Int J Urban Reg Res 23(4):670–696
37. Lin GCS, Yi F (2011) Urbanization of capital or capitalization on urban land? Land development and local public finance in urbanizing China. Urban Geogr 32(1):50–79
38. Liu Y (2018) Introduction to land use and rural sustainability in China. Land Use Policy 74:1–4
39. Liu Y, Fang F, Li Y (2014) Key issues of land use in China and implications for policy making. Land Use Policy 40:6–12
40. Liu Y, Li J, Yang Y (2018) Strategic adjustment of land use policy under the economic transformation. Land Use Policy 74:5–14
41. Liu Y, Li Y (2017) Revitalize the world's countryside. Nature 548(7667):275–277
42. Liu Y, Zhang Z, Zhou Y (2018) Efficiency of construction land allocation in China: an econometric analysis of panel data. Land Use Policy 74:261–272
43. Long H, Zou J, Pykett J, Li Y (2011) Analysis of rural transformation development in China since the turn of the new millennium. Appl Geogr 31(3):1094–1105
44. Ma LJC (2002) Urban transformation in China, 1949–2000: a review and research agenda. Environ Plan A 34(9):1545–1569
45. Minguillo D, Thelwall M (2015) Research excellence and university-industry collaboration in UK science parks. Res Evaluat 24(2):181–196
46. National Bureau of Statistics of China (2019) China statistical yearbook. China Statistics Press, Beijing [中国国家统计局. 2019. 中国统计年鉴. 北京: 中国统计出版社]
47. National Bureau of Statistics of China (2019) China city statistical yearbook. China Statistics Press, Beijing [中国国家统计局. 2019. 中国城市统计年鉴. 北京: 中国统计出版社]
48. Official Homepage of the Ministry of Science and Technology of the People's Republic of China (2019) Retrieved from www.most.gov.cn. Last accessed on 28 Dec 2019
49. Official Homepage of the National Bureau of Statistics of the People's Republic of China (2019) Retrieved from www.stats.gov.cn. Last accessed on 28 Dec 2019
50. Official Homepage of the State Council of the People's Republic of China (2019) Retrieved from www.gov.cn. Last accessed on 28 Dec 2019
51. Oh DS (2002) Technology-based regional development policy: case study of Taedok Science Town, Taejon Metropolitan City, Korea. Habitat Int 26(2):213–228
52. Phelps NA, Dawood SRS (2014) Untangling the spaces of high technology in Malaysia. Environ Plan C 32(5):896–915
53. Sargeson S (2013) Violence as development: land expropriation and China's urbanization. J Peasant Stud 40(6):1063–1085
54. Shen J, Wu F (2017) The suburb as a space of capital accumulation: the development of new towns in Shanghai, China. Antipode 49(3):761–780
55. State Council of the People's Republic of China (1988) Interim regulations of the Beijing experimental zone for the development of new technology industries. Retrieved from www.beijing.gov.cn/zfxxgk/110001/szfwj/1988-05/20/content_931610983654431bbe859a7bc16f18e2.shtml
56. Suzuki S (2004) Technopolis: science parks in Japan. Int J Technol Manage 28(3–6):582–601
57. Vaidyanathan G (2008) Technology parks in a developing country: the case of India. J Technol Transfer 33(3):285–299
58. Wang J, Leng T (2011) Production of space and space of production: high tech industrial parks in Beijing and Shanghai. Cross-Currents 1:100–126
59. Yang Y, Liu Y, Li Y, Du G (2018) Quantifying spatio-temporal patterns of urban expansion in Beijing during 1985–2013 with rural-urban development transformation. Land Use Policy 74:220–230

60. Ye C, Chen M, Chen R, Guo Z (2014) Multi-scalar separations: land use and production of space in Xianlin, a university town in Nanjing, China. Habitat Int 42:264–272
61. Ye C, Chen M, Duan J, Yang D (2017) Uneven development, urbanization and production of space in the middle-scale region based on the case of Jiangsu province, China. Habitat Int 66:106–116
62. Yeung HWC (2014) Governing the market in a globalizing era: developmental states, global production networks and inter-firm dynamics in East Asia. Rev Int Polit Econ 21(1):70–101
63. Yeung HWC (2016) Strategic coupling: East Asian industrial transformation in the new global economy. Cornell University Press, New York
64. Yeung HWC, Coe NM (2015) Toward a dynamic theory of global production networks. Econ Geogr 91(1):29–58
65. Yew CP (2012) Pseudo-urbanization? Competitive government behavior and urban sprawl in China. J Contemp China 21(74):281–298
66. Zeng G, Liefner I, Si Y (2011) The role of high-tech parks in China's regional economy: empirical evidence from the IC industry in the Zhangjiang High-Tech Park, Shanghai. Erdkunde 65(1):43–53
67. Zhang F, Wu F (2012) "Fostering indigenous innovation capacities": the development of biotechnology in Shanghai's Zhangjiang High-Tech Park. Urban Geogr 33(5):728–755
68. Zhao P (2015) The evolution of the urban planning system in contemporary China: an institutional approach. Int Dev Plan Rev 37(3):269–287
69. Zhao P, Li P (2017) Rethinking the relationship between urban development, local health and global sustainability. Curr Opin Environ Sustain 25:14–19
70. Zhuang L, Ye C (2018) Disorder or reorder? The spatial production of state-level new areas in China. Sustainability 10(10):3628
71. Zhuang L, Ye C, Ma W, Zhao B, Hu S (2019) Production of space and developmental logic of New Urban Districts in China. Acta Geographica Sinica 74(8):1548–1562 [庄良, 叶超, 马卫, 赵彪, 胡森林. 2019. 中国城镇化进程中新区的空间生产及其演化逻辑. 地理学报, 74(8): 1548–1562]
72. Zhuang L, Ye C, Lieske SN (2020) Intertwining globality and locality: bibliometric analysis based on the top geography annual conferences in America and China. Scientometrics 122(2):1075–1096

Chapter 7
Multi-scalar Separations: The Production of Space of University Towns in China

Abstract The university town is an important phenomenon in the course of urbanization in China. This article introduces and applies theory of production of space and socio-spatial dialectic to explain the processes and mechanisms of production of space in Xianlin university town in Nanjing City, China. As a typical case, Xianlin university town displays multi-scalar separations. The time-scale separation has four sides: the old and new campuses are completely different; teachers and students spend a significant amount of time commuting that they cannot communicate well; during summer and winter vacations, the university town becomes an "empty nest"; and life of low-income earners is fragmental. The four kinds of spatial scale splits are inside the campus, between universities and downtown, among universities, and between the city and its citizens. Resources and rights have an imbalanced distribution among the different classes, which leads to social space differentiation or alienation. The powers of discourses and land resource distribution are in the hands of the government. University managers are stimulated by the idea of a "larger and newer campus" and keep a watchful eye on competing for more land resources. Planners usually cater to such ideas. However, teachers, students, and low-income earners of the university town are neglected. Social process and the influences of land-use/cover change (LUCC) should be more frequently discussed in the future.

7.1 Introduction

Urbanization in China has drawn the attention of the world. A report from the United Nations Department of Economic and Social Affairs/Population Division [38] shows that Asia, especially China, will continue to urbanize more rapidly in the coming decades. Accordingly, more research and debates on the urbanization of China have been conducted. Most of these research and debates focus on population, institutions and policies, land use, ecological and environmental issues, relationship between

---

This chapter is based on [***Habitat International***, Ye, C., Chen, M., Chen, R., & Guo, Z. (2014). Multi-scalar separations: land use and production of space of Xianlin, a university town in Nanjing, China. *Habitat International*, 42, 264–272].

urbanization and industrialization or gross domestic product, and speed of urbanization, among others [1, 3, 4, 9, 27, 28, 33, 41]. Traditional themes are still studied until now. Aside from these issues, the university town, which is a new and recent type of urbanization in China [40], should also be given attention.

A university town is a highlighted phenomenon reflecting urbanization and LUCC in China. The "University town" first appeared in developed countries as a higher education phenomenon [10]. In Europe and the United States, many university towns have a long history and have gradually become "knowledge cities" [8, 14]. Globalization has caused the construction of China's university towns to be inevitably influenced by the styles of the United States, Japan, and some European countries, and to share the same characteristics with them. University towns in China have already become the way for officials to boost urbanization, thus, they have shown strong government-oriented characteristics that several scholars refer to them as "from top to bottom" mode [25, 40]. In the course of this kind of urbanization, a university town is usually built in areas that did not have universities before, rapidly transforming farmlands to educational or industrial land. University towns have sprung up only in the last 10 years or even later. In this regard, understanding the relations and interactions between the development of a university town, LUCC (land-use/cover change), and society is essential to discover how these significant changes occur in so short a time.

LUCC is an interdisciplinary field studied by scholars from different fields, such as natural and social scientists, planners, and geographers. However, according to the Global Land Project [11], most studies on LUCC mostly focused on the natural and ecological side. Moreover, the consequences on the social system are not yet fully understood, including the challenges to social justice and conflicts. Common sense dictates that people are at the LUCC core, although this idea seems easy to disregard. LUCC should be regarded as a generalized social process because society is composed of different people. We should choose several social science theories and introduce them into the study of LUCC to understand and explain this concept from the social view.

The theory of production of space is such a kind of social theory that provides us with a set of profound thought systems that closely connect society with space and time. Applying the new theory to LUCC research is necessary and interesting for a better understanding of LUCC as a social process. However, only a few articles have considered this theory for the study of LUCC because the theory is ambiguous. This paper attempts to overcome this challenge according to applying the theory of production of space to a typical and practical case analysis of university town development.

7.2 Theory and Framework

Theory of production of space is fundamental in Marxist geography or the neo-Marxist urban school [31, 34]. It first appeared in the 1970s and is often used to explain the urbanization of developed countries. However, Chinese scholars have begun to focus on this theory and put it into practice only since 2000 [40].

As a critical theory, production of space generally means that the urban landscapes and spatial structures have been reshaped by several political, economic, and social factors, mainly capital, power, and class, so that the urban space finally becomes their production and process. Before the 1970s, the concept of space was usually ignored in traditional philosophical ideas or epistemology, including classical Marxism, and was regarded as a physical or abstract factor that had no connection with social concepts [15, 17, 20, 22]. Based on a complete critical scrutiny of these kinds of opinions, Henri Lefebvre, a great French Marxist thinker and philosopher, originally presented the idea of "(social) space is (social) production" in his masterpiece Production of Space and held that the relationship between space and society should be rethought [22]. In the view of Lefebvre, social relations are spatial relations and vice versa, urbanization driven by capital dissimilates social (spatial relations and disregards the needs of vulnerable groups. Thereafter, his followers developed this theory and applied it to the discussion of urban issues and uneven geographical development [2, 12, 16, 18, 21, 30, 34, 36, 37, 39].

The concept of social space makes theory of production of space different from past and other theories. The concept of "social" in this theory should be considered in a broader sense, reflecting all social actions and relations. We can divide the "social" concept into three parts: political, economic, and (narrow) social, which correspond to power, capital, and class, respectively. The core of politics is power, and power (including discourse and knowledge) controls over society and produces a different space [5–7]. Capital is the most important factor in economic actions, as stressed by many thinkers from Karl Marx to David Harvey, and it flows and distributes in different areas so that it produces uneven space, such as the developed and developing regions [16, 19]. Class in society occupies a position similar to that of capital in the economy. In the course of urbanization, low-income earners and high-income earners have different spaces for living, and space is usually used to separate the poor or workers from the rich or capitalists [23, 24].

Space and society are no longer treated as two different ideas but as one idea or two sides of an idea because of the three forces [34]. Harvey [15] tried his best to integrate social process and spatial form with the concept of "social-process-spatial-form" as evidenced in several cases of urban development. Based on this idea, [35, 36] introduced the notion of socio-spatial dialectic, which emphasizes interactions and dialectic between time, space, and society, and makes the three aspects equally important and inseparable. According to socio-spatial dialectic, "social processes produce scales and scales affecting the operation of social processes. Social processes and space—and hence scales—mutually intersect, constitute, and rebound upon one another in an inseparable chain of determinations" [13]. The question then is as follows: How can we link this ambiguous theory to LUCC and policies?

This article designs a simple framework (Fig. 7.1) to solve the problem and show the relations between LUCC, policies, and production of space. Based on this framework, LUCC, policies, time, and production of space can be divided into three scales. In LUCC research, social, cultural, economic, and political influences should be

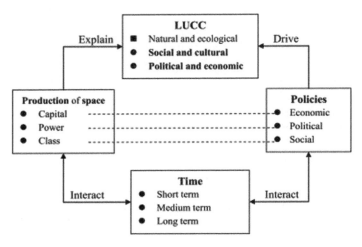

Fig. 7.1 Framework on the relations among LUCC, policies and production of space

considered aside from the natural and ecological effects. LUCC policies can be divided into economic, political, and social types, which respectively correspond to three factors (power, capital, and class) in the course of production of space. As previously mentioned, power, capital, and class also correspond to political, economic, and (narrow) social policies, respectively. Three kinds of time scales interact with LUCC policies. A different time scale has a different effect on LUCC and production of space. In general, spatial representations of LUCC reflect cultural heritage or turns in the long term (more than 10 years); more economic and political changes are reflected in the medium term; and social change or everyday life is reflected in the short term. Policies often change and drive LUCC, and theory of production of space can be used to explain the social, political, and economic processes of LUCC.

7.3 Xianlin University Town: Its Land Use and Policies

7.3.1 Context

Xianlin university town is a typical case in the construction of university towns in China. Three reasons explain why this is so.

First, Xianlin university town is located in the Qixia District of Nanjing City, Jiangsu Province, which is part of coastal China and one of the most developed regions in the country. Over 80% of university towns are distributed in this region (Fig. 7.2). With a long history and rich cultural accumulation, Nanjing is one of the best-known ancient capital cities in China and is the long-term capital of Jiangsu Province. Since the 3rd-century AD, 10 dynasties had been established in Nanjing,

7.3 Xianlin University Town: Its Land Use and Policies

including the epochal Republic of China founded by Sun Yat-sen. Nanjing is also an important central city in the middle and lower reaches of the Yangtze River and belongs to the northern subtropical monsoon climate zone.

Second, Xianlin university town has matured more quickly than the many other university towns in China. It was first planned in the early 1990s and built in 2002. Over a decade was spent from the beginning of planning; therefore, it has more experience and history, which can provide us with more valuable materials.

Third, Xianlin university town is one of the large-scale university towns in China. At the beginning of the planning stage, the designed area was 70 km^2, which is considered the largest in China. In 2011, its built-up area was 47 km^2, and its population numbered to 260,000. Its number of undergraduates and postgraduates was 120,000 from 12 universities and colleges, including Nanjing University and Nanjing Normal University (NJNU), accounting for 8% of the total number of undergraduates and postgraduates in Jiangsu Province ([29], see also Fig. 7.2).

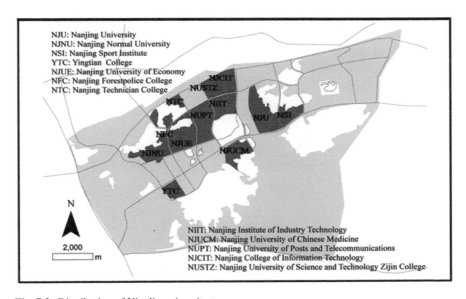

Fig. 7.2 Distribution of Xianlin university town

7.3.2 Xianlin University Town LUCC

From 2002 to 2012, the Xianlin university town LUCC was characterized by strong growth, cyclical fluctuation, and diverse types, as shown in Figs. 7.3 and 7.4. The land use types were mainly industrial, residential, educational (including university, primary, and high school), governmental, water, business, and undeveloped land. In general, Xianlin university town has not only realized a transition of land use types from merely a few university and residential lands to diverse land use, as previously mentioned, but has also marked an increase in total number and scales of land use. From a village or almost wilderness land, the Xianlin area has become more urbanized since 2002, and this trend will clearly continue. In particular, a significant increase in university lands, along with residential and industrial lands, indicates a developmental balance between universities, urban areas, and its industries.

The development of Xianlin has mainly experienced three stages since 2002 (Fig. 7.4 and Table 7.1), as seen from the changes of land use and its policy degree. In the first stage (2002–2004), Xianlin built several universities but built only a few other types of land use. The second stage (2005–2008) was the most turbulent period, although the residential and industrial lands were largely developed. In the third stage (2009–2012), LUCC gradually flattened every year. Residential lands were the lands that increased the most. Over time, Xianlin has become "like" a university town.

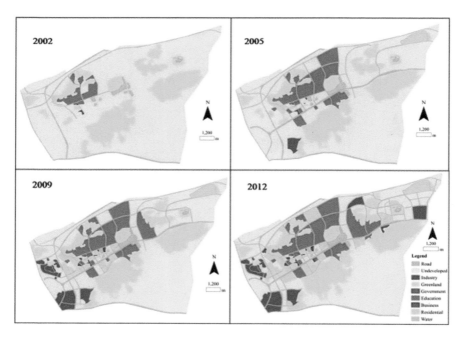

Fig. 7.3 LUCC of Xianlin university town from 2002 to 2012. *Note* Authors' survey and Official Homepage of Xianlin University Town (http://www.njxl.gov.cn/)

7.3 Xianlin University Town: Its Land Use and Policies

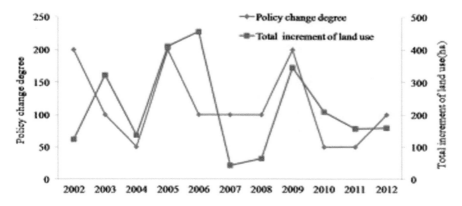

Fig. 7.4 Curve of change degree of policies and total increment of land use from 2002 to 2012. *Note* The policy change degree is equal to the influential degree of the big events (Table 7.1). The values of policy change degree are roughly assigned by authors based on the rank of authority figures inspected Xianlin and the size of impact on the directions of Xianlin University Town. Provincial official is assigned 100 and municipal official is assigned 50; if the developmental orientation of Xianlin University Town is fully changed, it is assigned 100; if the orientation is little changed, it is assigned less than 50. The "total increment of land use" is the increased area of developed land each year than in the previous year

Table 7.1 Big events and developmental orientations on Xianlin from 2002 to 2012

Year	Big events	Orientation	Key points
2002	University Town established; provincial Secretary Li said	A masterpiece	Supreme leader; beginning
2003	Secretary Luo of Nanjing said	First-rate university town in 5 years	
2005	Mayor of Nanjing inspected Xianlin	A new higher-level urban area	Functional pluralism: city, industry and university
2006	Secretary of Nanjing said	National first-rate University town	
2007	The national economy and social development goal of Nanjing	Focus on business and CBD	
2008	Mayor of Nanjing inspected Xianlin	Improvement by innovation and science	
2009	Administrative organ of Qixia district relocated in Xianlin	New development center of Qixia district	Transfer of local government
2012	Municipal government repositions Xianlin	A Science City	

Source Authors' survey and organize

This kind of unstable developmental pattern is the reason why we refer Xianlin as "like" but not "is" a university town. The word "like" implies that several potential risks need to be discovered and solved. In fact, the developmental instability of Xianlin university town is the most worrisome problem. Therefore, we must answer why and how such phenomenon could happen. Several special policies that have driven Xianlin LUCC should be considered first because they are different from the common policies used in LUCC analysis.

7.3.3 Policies and Their Influences on Xianlin Town's LUCC

The development of Xianlin has been affected by policies, especially instructions from the senior officials. From the start of the Xianlin construction planning to the present, many Nanjing City and Jiangsu Province leaders have frequently visited Xianlin and delivered instructions for the position of the area (Table 7.1). Based on the three main stages, the developmental orientation of Xianlin has undergone three main transformations: from a pure university town to a new center of the Qixia District and to one of three sub-cites of contemporary Nanjing. The years 2002, 2005, and 2009 were the turning points in Xianlin development (Table 7.1).

Policies that have been mainly formulated by senior leaders play an important role in the process, reflecting the strong influence of power. For example, in the entire development process (including the planning period in early 1990s), Xianlin has been receiving instructions from the supreme leaders of Jiangsu Province. Nevertheless, the developmental direction of Xianlin is not explicit, although the construction should clearly be a "masterpiece," as requested by the local leaders. Officials would repeat the same discourses, such as "first-rate," "higher level," and "scientific city." Another example is the transfer of local government departments in 2009. At that time, the administrative organ of Qixia district was relocated to Xianlin, providing the latter with a better chance to develop. The place where the administrative organ is located usually has significantly more rights and opportunities than others under the government-dominated model in China. A coincidence in the increment of residential, industrial, and university land use was not noted. The curve of policy change degree was determined to correspond to that of the total LUCC increment when the two curves were compared (Fig. 7.4 and Table 7.1). This result indicates that a far closer relation exists between policies and LUCC. However, Xianlin university town inevitably has experienced both progress and difficulties at least three times because the policies usually reflect only the vague and inconstant preferences and ideas of a few leaders.

7.4 Multi-scalar Separations of Xianlin University Town

Xianlin university town's LUCC can be explained from the three scales of time, space, and society based on the theory and methodology of socio-spatial dialectic. Serious separations exist within the Xianlin area.

7.4.1 Time-Scale Separations

The four kinds of time-scale separations are the old and new campuses; the more and less communicating time; the "empty" university town during summer and winter vacations; and the fragmental life of low-income earners.

The most remarkable and obvious differences and splits are indicated by the old and new campuses. On one hand, designers or planners of the new campus usually attempt to reconstruct a newer, larger, and more "modern" university to meet the tastes of decision makers and managers, to forget the connections between the different campuses, and to avoid inheriting from the old. On the other hand, duplicating something from the old campus, even if planners wish to do so, is impossible because of the many historical landmarks. These landmarks include structures built during the Republic of China period and the former residence of Rabe in the old campus, which cannot be physically transferred or copied. Moreover, the operation of the new campus has resulted in a number of teaching and research facilities to be removed. The loss of various functions has extremely weakened and caused the gradual decline of the university. For example, Suiyuan is the old campus of Nanjing Normal University, designed by the renowned American architect Henry Murphy and his Chinese student Yanzhi Lu. It combines Western and Chinese architectural styles. It was recognized as a classic work during the Republic of China period and hailed as "the most beautiful campus of the East." By contrast, the new Xianlin campus of Nanjing Normal University pursued a different kind of style called modernization and internalization to be selected in the "211 Project," which refers to the selection of 100 excellent universities for inclusion in the world-level universities in the twenty-first century in China. The project was initiated by the Ministry of Education of the People's Republic of China in 1995 with the intent of raising the research standards of high-level universities [26]. The spatial forms and landscapes of the new campus usually have no connection with the old. The same problem has happened to the reconstruction of Suiyuan campus in the Xianlin campus (Fig. 7.5).

Teachers and students spend a significant amount of time commuting that they cannot communicate well. Many teachers still live in the central area of Nanjing, although universities have been moved to Xianlin and most of the teaching activities are also located there. Nanjing is a long distance from Xianlin. The difficult travel by bus takes one hour or more, which inevitably affects teaching work and the communication of teachers with their students. During vacations, especially winter vacation, the university town becomes an "empty nest" because most of the students

Fig. 7.5 Some representations of time-scale and space-scale separations. *Source* Authors' survey and photo. *Note* (a) Suiyuan campus of NJNU built in Republic of China period; (b) Xianlin campus of NJNU built in 1998; (c) Gulou campus of NJU built in Republic of China period; (d) Xianlin campus of NJU built in 2009; (e) Empty schoolyard of NJUE in summer vocation; (f) Long queue waiting for school bus in NJNU; (g) Empty shopping mall of Dacheng in summer vocation; (h) Steep slope not suitable for walking in NJNU

return home. The "empty nest" of the university town results from the government ignoring low-level businesses. Warehouse stores, thrift shops, and shopping malls (e.g., Dacheng and Aishang shopping malls in Xianlin), which meet the various consumer demands of students, almost do not have visitors. Shopkeepers have to suspend their businesses during vacations.

Life of low-income earners is fragmental. They work and commute for a significant period so they do not have time for daily leisure. The Xianlin university town management committee spends over 20 months in resettling more than 14,000 households and more than 40,000 residents in the outskirts of Xianlin [29]. However, jobs provided by the government for them are far from enough. Those who lose their land become unemployed because of the lack of training and knowledge base for a well-paid job. Living in the peripheral area has also caused laborers to spend at least two hours commuting every day, making their lives fragmentized.

7.4.2 Space-Scale Separations

Space, society, and time are indivisible and interactive based on the socio-spatial dialectic and frame of this paper. Time-scale separations necessarily involve socio-spatial factors and vice versa. Although Fig. 7.5 mostly shows the changes in time scale, different spatial forms are an intuitive perception. The four kinds of spatial separations are inside campus, between universities, between universities and downtown, and between the city and the citizens.

The new campus can obtain significantly more adequate land resources than the old campus because the former is located in the suburbs of Nanjing City. Campuses are too large to do anything by foot, and the commuter buses inside the campus are limited and have a long interval time. Sometimes, especially during rush hour, buses are so crowded that students and teachers are packed similar to a can of sardines. As one of the authors of this paper and a teacher at NJNU, I have been in such situations several times. A long queue of people waiting for the bus is a common sight. Walking is an exhausting endeavor when going from one place surrounded by mountains to another, especially when the only link from place to place is a flyover. The connections between one place and another in the campus are separated over time.

Moreover, only a few connections exist between universities. At the start of the project, one of the important aims of the university town construction was to build a university union. As early as 2004, under the auspices of the administrative committee, nine universities in Xianlin planned to establish a "teaching union," which attracted widespread attention at that time but has made no substantive progress until now. Using agglomeration advantages to promote exchange and sharing of resources among universities has become lip service.

University towns in other countries, such as Brown in the United States and Wageningen in the Netherlands, have successfully connected universities with their local towns. Xianlin university town in China has yet to connect with the local town. The well-known term "town and gown" refers to conflicts between citizens and students in Oxford University town because of different religions and habits. Attracting teachers and students to several commercial centers planned by the Xianlin Management Committee is difficult. The Golden Eagle Group, a famous upscale corporation in Nanjing, built a shopping center in Xianlin in 2009 but it failed to become popular. This shopping center eventually became a discount store. Based on the planning of the Xianlin Management Committee, three more commercial centers will be built in Xianlin in the next several years. Nevertheless, college students are unwilling to spend their money there. "Usually, we do not go there to buy. Although I'm not an indoors man, I really don't want to go there because the clothes sold in Xianlin are very expensive and do not have many choices. The products are not as good as those sold in TaoBao Electronics Store. If we really want to go shopping, we go downtown," an undergraduate from Nanjing Normal University told the authors.

7.4.3 Social Separations

The "social" concept is important in theory of production of space. However, "society" is so complicated and abstract that it cannot be completely understood. The concept will be easily understood if we grasp its core meaning: class or community. The different classes in Xianlin university town are government officials, planners, university administrators, teachers and students, residents (including the low-income and high-income earners), corporations, and college and university unions (Figs. 7.6 and 7.7). A clear differentiation or alienation between several classes exists because power and rights are distributed in an off-balance manner among the different classes.

Fig. 7.6 Distribution of illegal inns and cars in Xianlin university town. *Source* Authors' survey and photo. *Note* (a) There is a sign "Cherish your life, and stay away from illegal cars" (in the red ellipse) on the door of NJUE, but many illegal cars are just outside. (b) An illegal inn earner tries to attract passers to rent her rooms. (For interpretation of the references to color in this figure legend, the reader is referred to the web version of this article.)

Along with the promotion of urbanization and the development of the university town, the number of suburban peasants confronted with land expropriation and housing removal has increased in recent years. Farmers who lost lands have formally become town dwellers. However, they have difficulty finding suitable and decent jobs because of their low knowledge base and the lack of enough government care. Thus, a number of farmers have to do heavy or strenuous manual work in universities, such as being cooks, service personnel, or cleaners. For them, these jobs normally consume longer time, leave less leisure time for them, and provide little pay and minimal rewards. Others who are unwilling to do such works prefer to be engaged in certain non-traditional jobs such as illegal inn earners and street vendors (Fig. 7.6). For example, the limited public transport system has resulted in a flourishing illegal car business. Ironically, a number of universities usually have signs that say, "Cherish your life, and stay away from illegal cars" (in the red ellipse), but illegal cars remain in the vicinity of the university. Illegal car drivers have already damaged the job prospects of taxi drivers and have caused frequent traffic accidents. Conflicts have also occurred between vendors and urban inspectors as well as between inn earners and students.

Dealing with these problems has become a huge challenge for university town managers. More seriously, a significant social differentiation and gap exist between

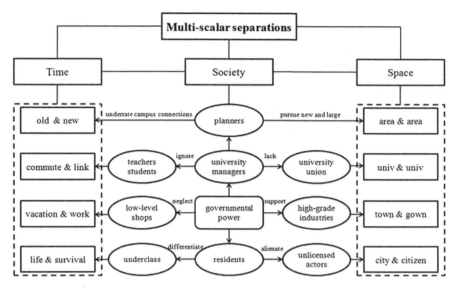

Fig. 7.7 Mechanism of production of space in Xianlin

low-income earners and high-level classes. Low-income earners go to cheap stalls to shop and unlicensed inns for accommodation to cut down on expenditures and save more money. Illegal car drivers cluster near universities and subways (Fig. 7.6). The term "Xianlin new village" has yet to be formally introduced, but it is already an old idea that refers to a clustered area of low-income earners. However, the downtown center is now the location of a number of senior communities for the rich. Thus, the differentiation of social space is significantly prominent.

7.5 Production of Space Mechanism in Xianlin University Town

The production of space mechanism in Xianlin university town can be briefly summarized with the aid of socio-spatial dialectic. This study makes abstract theory understood more clearly and easily.

As shown in Fig. 7.7, the space of production in Xianlin is represented by multi-scalar (i.e., time, space, and society) separations. Each large-scale separation can be divided into four sub-scales, that is, space-scale separations in the order of small- to large-scale, which are from inside campus to interuniversity, to university and city, to city, and to its citizens. Why citizens are regarded as a spatial scale requires an explanation. Based on theory of production of space and postmodern geographies, the term "body" is the basic spatial unit. Significantly more studies have focused on body and urban space because a city space is built and developed by the citizen body

[5, 32]. The word "citizens" in Fig. 7.7 can also be treated as an abbreviation for "citizens' space."

The generalized concept of "society" emphasized by Lefebvre is located in the core parts of this figure and can be partitioned into several main classes or communities. In the societal scale, power is a keyword and it dominates the production of space to some extent. Local governments master the power of discourse and land resource distribution. After the struggle for attaining more lands, managers of universities want to have "larger and newer campuses" but ignore the unions, teachers, and students who are the true bodies in a university town. Planners commonly cater to these officials because of their reliance on the latter. However, connections between the old and new campuses and between the different areas in the campus are underrated. As for the residents, low-income people are so neglected that they have to work in illegal jobs.

Different spatial types belong to different classes, and different classes produce and consume these spaces. In this way, the Xianlin university town of Nanjing City displays multi-scale separations.

7.6 Discussion and Conclusions

Space of production or socio-spatial dialectic focuses on the interactions between time, space, and society of different levels. However, these kinds of interactions are significantly complex and enough to puzzle researchers. Fortunately, several specific and microscopic cases can be used for understanding and explaining such interactions. This article introduces and attempts to apply socio-spatial dialectic to the analysis of LUCC on the findings of several interesting and significant ideas. The most important idea is that multi-scalar separations clearly exist in Xianlin university town. Although the different university towns in China have several differences, many university towns have the same characteristics and dynamics. As far as Xianlin university town is concerned, its rise and fall are closely associated with government power. Other university towns in China seem to have the same features because of the government-oriented pattern. Therefore, the production of space in Xianlin university town is representative of what occurs in the country.

A university is a place where knowledge is produced, but this production cannot be divorced from the specific context of time, space, and society. The development of a university town is a complicated, dialectical, and interactive process between time, space, and society. It is related to urbanization and LUCC, with social space and process as the major segments. The process of rapid urbanization and rural–urban transformation has brought about many phenomena and problems about the production of urban space in China since the 1990s. As a new type of urbanization, the university town in China is largely different from that in the other countries. This article tries to discover such paradox based on the case of Xianlin university town: the most important aim of urbanization is to create free space and enough lands for those pursuing a better life, although most of them cannot share such space. However, they are thrown out from the space that they created themselves.

Therefore, we must present an important and much-debated question: Can we hold a kind of neutral value in studying LUCC? The answer is clearly "no", because the LUCC process appears to involve value judgments throughout. The subject and objectives of LUCC research are humans and their influences are on land; detaching one from the other is difficult. The researchers come from many different fields, and hold different values and attitudes even to the same LUCC problem. Many social sciences and humanities take emphasis on the role of the value judgments, and they can provide some inspirations and methods for LUCC research as an interdisciplinary field. Except for this case, LUCC should be regarded as a social process that must contain value judgments. For some developing countries like China, their social problems on land use are becoming more prominent than ever. Therefore, it is very important for scholars to judge justification of policies on LUCC, understand the relations between LUCC and social classes, and sympathize with vulnerable groups such as land-losing peasants, low-income earners. We cannot avoid discussing the value in LUCC research or veiling it similar to traditional positive analysis. We should actively promote interdisciplinary integration to study effectively the LUCC social process and its influences.

References

1. Bloom DE, Canning D, Fink G (2008) Urbanization and the wealth of nations. Science 319(5864):772–775
2. Castells M (1977) The urban question (Trans: Sheridan A). The MIT Press, Cambridge
3. Chan KW (2010) Fundamentals of China's urbanization and policy. China Rev 10(1):63–93
4. Cohen B (2006) Urbanization in developing countries: current trends, future projections, and key challenges for sustainability. Technol Soc 28(1–2):63–80
5. Foucault M (1977) Discipline and punishment. Tavistock, London
6. Foucault M (1980) Power/knowledge. Harvester, Brighton
7. Foucault M (1986) Of other spaces. Diacritics 16(1):22–27
8. Franz P (2008) From university town to knowledge city: strategies and regulatory hurdles in Germany. In: Yigitcanlar T, Velibeyoglu K, Baum S (eds) Knowledge-based urban development: planning and applications in the information era. Information Science Reference, New York
9. Friedmann J (2006) Four theses in the study of China's urbanization. Int J Urban Reg Res 30(2):440–451
10. Gilbert W (1961) The university town in England and West Germany. University of Chicago, Chicago
11. Global Land Project (GLP) (2005) Global land project: science plan and implementation strategy. IGBP Secretariat, Stockholm
12. Gottdiener M (1985) The social production of urban space. University of Texas Press, Austin
13. Gregory D, Johnson R, Pratt G, Watts M, Whatmore S (2009) The dictionary of human geography, 5th edn. Blackwell, Oxford
14. Gumprecht B (2003) The American college town. Geograph Rev 93(1):51–80
15. Harvey D (1973) Social justice and the city. Edward Arnold, London
16. Harvey D (1982) The limits to capital. Blackwell, Oxford

17. Harvey D (1985) The urbanization of capital. Blackwell, Oxford
18. Harvey D (1996) Justice, nature and the geography of difference. Blackwell, Cambridge
19. Harvey D (2000) Space of hope. Edinburgh University Press, Edinburgh
20. Harvey D (2001) Spaces of capital: towards a critical geography. Edinburgh University Press, Edinburgh
21. Harvey D (2010) The enigma of capital and the crises of capitalism. Profile Books, London
22. Lefebvre H (1991) The production of space (Trans: Smith N). Blackwell, Oxford
23. Lefebvre H (1996) Writings on cities. Blackwell, Oxford
24. Lefebvre H (2003) The urban revolution. University of Minnesota Press, Minneapolis
25. Li J, Mi Y, Yao S (2010) University town: Emerging urban area in Chinese urbanization process. China Science and Technology University Press, Hefei. [李峻峰, 米岩军, 姚士谋. 2010. 大学城:我国城市化进程中的新型城市空间. 合肥: 中国科学技术大学出版社.]
26. Li L (2004) China's higher education reform 1998–2003: a summary. Asia Pac Educ Rev 5(1):14–22
27. Lin GCS (2007) Chinese urbanism in question: state, society, and the reproduction of urban spaces. Urban Geogr 28(1):7–29
28. Long H, Li Y, Liu Y, Woods M, Zou J (2012) Accelerated restructuring in rural China fueled by 'increasing vs. decreasing balance' land-use policy for dealing with hollowed villages. Land Use Policy 29(1):11–22
29. Official Homepage of Xianlin University Town (2012) Retrieved from http://www.njxl.gov.cn/www/njxl/2011/zhzc.htm
30. Olds K (1995) Globalization and the production of new urban spaces: Pacific Rim megaprojects in the late 20th century. Environ Plan A 27(11):1713–1743
31. Quaini M (1982) Geography and Marxism. Blackwell, Oxford
32. Sennett R (1994) Flesh and stone: the body and the city in western civilization. W. W. Norton & Company, London
33. Shen J (2005) Space, scale and the state: reorganizing urban space in China. In: Ma LJC, Wu F (eds) Restructuring the Chinese city: changing society, economy and space. Routledge, London
34. Smith N (1984) Uneven development: nature, capital and the production of space. Blackwell, Oxford
35. Soja EW (1980) The socio-spatial dialectic. Ann Assoc Am Geogr 70(2):207–225
36. Soja EW (1989) Postmodern geographies: the reassertion of space in critical social theory. Verso, London
37. Soja EW (1996) Third space: journeys to Los Angeles and other real-and-imagined places. Blackwell, Oxford
38. United Nations Department of Economic and Social Affairs (UNDESA) (2012) World urbanization prospects: the 2011 revision. United Nations, New York
39. Unwin T (2000) A waste of space? Towards a critique of the social production of space. Trans Inst Br Geogr 25(1):11–29
40. Yang Y (2009) China's rapid urbanization and the appearance of the "University Cities": a political economy perspective. 21st Century 1:104–113
41. Zhu J (2004) From land use right to land development right: institutional change in China's urban development. Urban Stud 41(7):1249–1267

Chapter 8
The Lost Countryside: Spatial Production of Villages in China

Abstract Rapid urbanization, especially in many developing countries, is accompanied by the decline of rural populations and rural culture. Tangwan village in Shanghai in China is a typical case. Based on the theory of production of space, this article presents spatial production of rural culture from three dimensions: ideological space, superficial space and everyday life space, and analyzes the reasons for the gradual decline of rural culture. The development of rural culture in China is seriously affected by the national and local policies. The formal power of a place which is mainly represented by the administrative level, dominates many cultural changes. With the decline of Tangwan's position to a village in an urbanized Shanghai, its local rural culture is declining. Under the pattern of production of space dominated by power and capital the village is challenged to maintain its rural culture and to develop a new culture. The traditional rural culture cannot be protected; however, the new rural culture has not matured. The lack of everyday life space for village residents to express and develop culture is partly to blame. The culture is the root of the rural development. The lost culture leads to the lost village. Adequate everyday life space is important for rural culture to flourish and though those spaces need careful planning to accommodate and grow local cultural needs.

8.1 Introduction

The dynamic relationship between economic growth and spatial change is an important topic of international research [5, 48]. Since the reform and opening-up, China's economy has been growing rapidly, and great changes have taken place in rural and urban areas [44]. China's urbanization is a complex system involving population, economy, society, space and other aspects [10]. This complexity requires investigation from cross-disciplinary methods. Production of space is an important theory to explain the interaction between society and space [2]. The theory has played an

---

This chapter is based on [*Habitat International*, Ye, C., Ma, X., Gao, Y., & Johnson, L. (2020). The lost countryside: spatial production of rural culture in Tangwan Village in Shanghai. *Habitat International*, 98, 102137].

important role in urbanization research since the idea was introduced into China [33, 50–52].

While China's urbanization has brought tremendous positive changes, it has also widened the gap between urban and rural areas [53], bringing significant impact to rural areas [6]. The urban areas attract rural youth with abundant capital and employment opportunities, which leads to large-scale population migration and ageing rural communities [7, 24, 56]. As a consequence, urban fringe villages (*cheng bian cun*) are formed around the city, and urban villages (*cheng zhong cun*) are formed within the city [19], while the "hollow villages" and the aged population remain in the countryside due to population change [25]. Rapid urbanization encroaches on rural land, resulting in loss of productive rural cropland [14, 42]. Rural land use has undergone a drastic change as well as increased land pollution from the externalities of urban development [18, 24, 28, 55]. China's per capita cultivated land area has been lower than the world average, and the quality of cultivated land cannot be guaranteed [3, 9].

With the emergence of these rural problems, many scholars have begun to pay attention to rural China [16, 21, 35]. Promoting rural sustainable development is the essence of solving rural problems, and it is necessary to study rural problems in the whole regional system and from multiple perspectives, using interdisciplinary methods [23, 45–47]. On the comprehensive level, the coordinated development of urban and rural areas is the ultimate goal and rural space is the foundation for realizing integrated urban and rural development [27, 31]. Narrowing the income gap between urban and rural areas, adjusting rural structure and rural resource allocation are the main strategies for achieving integrated and fair urban and rural development [29, 30, 36, 41, 55]. At the same time, many scholars also focus on temporal changes in rural settlements and rural education [15, 37, 43, 49].

In the study of rural issues, culture is often overlooked. However, culture plays an important role in shaping community [1]. Some have discussed rural culture and believed that culture is significant [12, 13, 17, 38]. With the promotion of globalization, rural culture is greatly changed [22]. Unfortunately, traditional rural culture is at risk of being ignored, destroyed and replaced by urban culture and modern culture. In China, rural governance has been an essential tool to harness the rural population and achieve desirable political schemes [4]. However, the development of rural culture is seriously affected by national and local policies, and cultural development is confronted with unprecedented challenges from rapid urbanization [22]. Although some researchers investigate the role of capital and power and the influence of special ideologies and events on people's everyday life in rural cultural development [8, 11], there is little research into rural cultural development with the theory of spatial production. In this research, we design a new framework to explain the relationship between rural culture and spatial production based on a case study of Tangwan village in Shanghai, China.

This paper consists of five parts. Following the introduction, the second part introduces the research area and methodology. The third part indicates the process of rural cultural construction based on the case of Tangwan, and analyzes the reasons leading to the gradual decline of rural culture in that place. The fourth part discusses

8.3 The Lost Countryside

8.3.1 Ideological Space

8.3.1.1 The Neglected Rural Public Space and Unbalanced Living Space

Bottom-up rural construction plays a positive role in promoting rural development [26]. However, the rural construction in Tangwan village has become a platform for the government to complete the planning and construction goals and continuously ignores the needs of residents. In the "Construction and acceptance criteria of beautiful countryside in Minhang District" [39], a stipulation on the construction of "livable" beautiful countryside parking lot said: "The current construction land should be transformed into a parking lot on the basis of meeting the needs of the villagers."

However, during the investigation, it was found that the parking lot in Tangwan village was transformed from a large vacant space. The public space is located along the Yutang River, with a beautiful landscape and good environment, which is a preferred place to enrich residents' life and build local culture. For example, it can be a place for residents' fitness square or for regular cultural events. In addition, the parking lot is located on "T" road, which is supposed to facilitate the passage of vehicles. But in fact, the road along the Yutang River is so narrow that even if it is located on "T" road, it is not convenient for vehicles to pass (Fig. 8.3). Such planning does not meet the needs of residents or the transportation system. Moreover, the original public parking lot has not been properly rebuilt to meet the planning requirements. It looks temporary and untidy.

During the beautiful countryside construction, the regulation of river in the village and the environmental regulation of the front and rear of the house are important projects in the construction, which are committed to maintaining the appearance of the village and improving the living environment of the villagers. However, the function of power and capital makes the residents in an unbalanced living environment. Judging from the built environment, Tangwan village mainly has two types of housing: traditional dwellings and newly built dwellings. As for the traditional dwellings, the government only painted the walls to complete the "face project" to achieve the goal of rural construction. In order to pursue a modern lifestyle, families with superior private resources have built new buildings one after another, which forms a sharp contrast with the traditional dwellings in terms of style and living conditions (Fig. 8.4). The differentiated treatment in various aspects not only creates an unbalanced living environment, but also weakens the residents' sense of community.

Fig. 8.3 Parking space in Tangwan village. *Source* Authors' photos

Fig. 8.4 Unbalanced living environment in Tangwan village. *Note* The picture on the left represents the traditional dwellings of Tangwan village, and the picture on the right represents the newly built dwellings. *Source* Authors' photos

8.3.1.2 The Excluded Traditional Rural Culture

In the process of the constructing beautiful countryside in Wujing town, ecological villages have added multiple indicators of cultural construction compared with other two types of villages. Tangwan village has a profound cultural heritage like Tiger dance listed as the intangible cultural heritage of Minhang District. The house of Peng's Garden (Fig. 8.5) is a cultural relic protected in Minhang District and was included in the immovable list of the third national cultural relics census of Shanghai on June 6, 2012. However, Tangwan village was excluded from "ecological" village status which is the only type that has cultural construction content. This effectively diminishes Tangwan's cultural construction and cultural protection. The original "street culture" of Tangwan Street has been replaced by the popular fashion culture.

8.3 The Lost Countryside

Fig. 8.5 Peng's Garden house. *Source* http://blog.sina.com.cn/s/blog_5accebe30102y4cn.html

Peng's Garden house gradually disappeared from the sight of local residents due to the prohibition from entering. Traditional cultures such as tiger dance have also been gradually forgotten. In an interview, a village committee director said:

> There is no tiger dance in Tangwan now. It used to belong to Tangwan Town, and now it does not exist in Wujing Town.

The report of the Shanghai Urban Master Plan (2017–2035) mentioned that it is necessary to strengthen the protection of the intangible cultural heritage that represents Shanghai's local culture and the protection of various traditional cultures closely related to people's quality of life. In the protection of historical towns, it is mentioned that towns with great historical and memorial significance, or towns that can reflect the traditional style and features of a certain historical period and the regional characteristics of Shanghai will be protected. However, as far as Tangwan culture is concerned, it is on the edge of defining "representative immaterial culture" and "traditional culture closely related to people's life". Tangwan village, once the center of Tangwan people's commune and Tangwan Township as well as the seat of the government, is actually an important area reflecting the developmental history of the previous Shanghai County and the current Minhang District. But now, it does not accord with the protection scope of "villages with distinctive features", nor does it accord with the protection scope of "historic towns" so its history and culture are not protected by the State. With the decline of its administrative level, the government has paid less attention to it. Therefore, its culture is not reasonably protected.

With the national strategy of Rural Revitalization and the Construction of Beautiful Countryside, local governments are scrambling to classify villages. As a village-level administration, Tangwan is hindered in the development of rural culture. Places with good location cannot be used as public places for cultural activities. The natural environment and housing conditions in Tangwan are separated and the contrast is strong. There are conflicts between cultural protection and cultural inheritance, and protection is often more important than inheritance. Traditional architecture cannot be visited, making it difficult for locals and outsiders to value and memorialize. The cultural development of Tangwan village is gradually declining.

8.3.2 Superficial Space: Culture Only on the Wall

Posters, exhibition boards and other exhibition methods are still the most mainstream way to publicize community culture in Tangwan village, which has become a special spatial form. The original Tangwan Street was once a lively place for people to gather and trade. It was the political, economic and cultural center of the Tangwan area. Walking into Tangwan village today, an explanatory panel about "memory of Tangwan Street" will be seen (Fig. 8.6). In the exhibition board, ancient house tiles are used to shape the appearance of ancient buildings. The content is divided into three parts: Tangwan Street, local celebrities and house buildings. However, each part is a very basic and simple introduction. The space around the exhibition board is surrounded by iron pillars to prevent the cars from covering the content of the exhibition board, which shows the local people's desire to see the exhibition board. Therefore, the exhibition board becomes a space of contradiction. On the one hand, it shows its past and expresses the memory. On the other hand, it uses simple content and "formalism" to show history and culture.

The wall of Tangwan village along the street and the bulletin board in the square are publicity posters for publicizing the village regulations, the creation of a national civilized city and other public welfare posters (Fig. 8.6). Undoubtedly, these are one of the contents of the community culture, which can foster community awareness and resident responsibility. However, the content does not express those closely related to residents' life, such as the publicity of community cultural activities, local news announcements and exhibitions, that is, it lacks local cultural characteristics. The village rules and regulations formulated by the Tangwan village committee are basically same as those of other villages. Moreover, by comparing the regulations of Tangwan village with those of Heping village, we found that the regulations of Heping village are more detailed, which not only involves traditional virtues, spiritual civilization and forbidden behaviors, but also elaborates the registration of migrant households and various behaviors that affect the environment and safety. The role of village regulations in "beautiful countryside" construction in "construction and

8.3 The Lost Countryside

Fig. 8.6 Some representations of superficial space. *Note* **a** "Memory of Tangwan Old Street" panel; **b** The picture of Tangwan Old Street; **c** Superficial village rules and regulations of Tangwan; **d** Bulletin board on safe community. *Source* Authors' photos

acceptance criteria of beautiful countryside in Minhang District" is defined as: "establish and improve the responsibility and restraint mechanisms for villagers to jointly protect the environment ", but such rules make village regulations almost the same for each village, which seems to be contrary to the content of "focusing on the local characteristics and reflecting local customs" mentioned in the strategy of rural revitalization. In addition, the poster establishing the national civilized urban area shows that the construction of Tangwan village is part of the construction of the Minhang urban area. Therefore, the community construction of Tangwan village is increasingly inclined to urban construction, and the rural flavor is not well represented.

8.3.3 Everyday Life Space: Limited Organization and Unformed Culture

The daily community activities of residents in Tangwan village mainly include listening to stories, playing board games, reading, watching movies and dancing. The daily activity space is mainly divided into two types: indoor activity center and outdoor fitness square. The former is mainly organized activity space set up by government departments and oriented to the elderly groups. The latter is a new activity space spontaneously formed by residents in the original public space.

Fig. 8.7 Some representations of everyday life space. *Note* **a** Listen to stories in the elderly activity center; **b** Neighborhood Center of Hetang Court; **c** Artistic activities in the fitness square of Sanxin Street; **d** Artistic activities on the long flower gallery.*Source* Authors' photos

There are two main residential activity centers in Tangwan. One is the Tangwan elderly activities center, which is located at the intersection of Beiwu road and Sanxin Street. It is a relatively small and old facility near the Tangwan vegetable market. The main content of the activity is to provide listening and drinking tea for the elderly at the center at noon every day (Fig. 8.7). The time of each activity is about one and a half hour, but each story will be told for about a month. There will be special staff to pour hot water for the elderly. There are a lot of people listening to books in the library (mainly male elderly).

The second is the Neighborhood Center of Hetang Court. In 2008, due to the inconvenience of some elderly people to go to the Tangwan activity center, the government built the Neighborhood Center of Hetang Court at the southern end of Hetang Bridge. The Neighborhood Center has a two-storey building and more than 500 m^2 of fitness space, 11 rooms with 8 different functions of the activity room, including audio-visual room, dance room, chess room, health room, leisure area, chat bar, happiness club and micro book bar. Compared with the old Tangwan activity center, the activity space provided here is more abundant, the types of activities are more diverse, the facilities are more complete and the decoration is more exquisite. Although the activity space is called neighborhood center, it is still aimed at the elderly group. The activities of the center will last from morning to afternoon. Generally, in the morning, the elderly people dance; at noon, a film or publicity film will be shown, and in the afternoon, many old people come to the chess room to play mahjong or cards. In addition, the neighborhood committee will hold an annual performance in Hetang court, and cooperate with the Minhang District government to carry out some community cultural activities, such as the "Happy Neighborhood" series of fun competitions.

The outdoor activity space is mainly a fitness square located in Sanxin Street (Fig. 8.7). In the morning, many people go to dance or exercise. In normal times, there are elderly people who gather spontaneously to play and sing, including blowing Suona, pulling Erhu and singing Huangmei opera. During the field investigation, an old man was observed to be blowing Suona and another was pulling Erhu. The songs they played were selected from some classic Huangmei operas in the 1950s and 1960s, with other classical movie episodes. During the interview, the old man who blew his Suona earnestly asked us to help them publicize this traditional music in schools, which showed the old generation's love of traditional culture.

The director of the village committee also mentioned the lack of funds for community activities many times in our interview. For example, he said:

> The reading club was originally proposed by the government to run, and they provided 100 thousand Yuan of construction funds. At the beginning, it was totally 2 Yuan for drinking tea and listening to stories. Later, when the leaders inspected, they said that we cannot take the money, but this is unrealistic, so we changed to take 1 Yuan, however, we also have to pay 350 Yuan to the storyteller.

He also mentioned that the neighborhood center had no follow-up funding after the initial funding:

> If we follow the standardization management, we need to send three people to manage and clean up, and a monthly wage will add up to 7000 Yuan; In addition, tap water is 1000 Yuan a month; water and electricity costs 2000 Yuan a month and so on, these follow-up costs are quite large, but without policy subsidies, we need to take our own money to hold activities. In short, if we refine these policies, it will be very difficult.

Regarding other cultural activities, such as dancing, he also said as below:

> The village committee will organize three or four dancing competitions every year, but due to the lack of funds, it is impossible to organize more dancing activities and can only be carried out simple activities. Yet, residents usually organize and choreograph dances by themselves.

8.4 Conclusions and Discussions

The transformation and disappearance of culture are a very complex problem. According to the case of Tangwan village in Shanghai, this article presents the development and reconstruction of rural culture through different spatial forms, and attempts to explain the interaction between power and spatial changes using the theory of production of space. In the process of rural construction in China, rural culture is closely related to power, which has become the leading force of cultural development and has penetrated different cultural spaces. Moreover, the government-led mode of cultural development has dominated rural reconstruction in China.

Lefebvre [20] once divided the production of space into three spatial levels: spatial practices, representations of space and the practice of space. According to the development of Chinese countryside, the production of rural space can be redefined as three types: ideological space, superficial space and everyday life space. Everyday life space is affected by the combination of ideological space and superficial space. However, the ideological space lacks the public space for cultural exchange and the superficial space is just a form. The spontaneous activities of residents form a new cultural space that resists the power and capital to rebuild public space and carry out cultural activities.

The production of cultural space in Tangwan village is manifested in three dimensions. In the ideological space, because of the decline of Tangwan's administrative level, Tangwan village is not only built with a lower standard than other villages, but also the traditional culture is disappeared. In addition, rural construction ignores the real needs of residents and becomes a platform for local governments to accomplish construction goals to show their political achievements. The difference in capital also leads to the obvious differences in the residents' living environment as well as the decline of residents' local identity.

As for superficial space, the exhibition board is one of the best places for local culture display, while the officials did not take specific measures to promote local culture, and the formalized similar writing has been widely different from the original intention of rural culture construction. In the everyday life space, the spontaneous activities of local cultural development are significantly larger than those organized activities. Due to insufficient funds, many cultural spaces are greatly hindered. Reconstructing rural culture has become the concern of rural revitalization.

The protection and inheritance of rural culture require the joint efforts of all sectors of society. The government should focus more on cultural construction and meet the needs of residents through multiple channels. As the holder and inheritor of rural culture, villagers can promote the rural culture in various ways. The rural areas should constantly update the existing culture and rebuild the cultural ecology so as to remake a place-based culture. Facing with the strong government, the villagers themselves should "voices" [26]. For example, the residents can spontaneously organize some cultural publicity activities and carry out more non-investment cultural construction activities. Besides, the participation of others will be a catalyst for cultural development. Planners should follow the basic principle of meeting the needs of residents and plan the rural public space to express local cultural needs. The community could bring culture to campus and let students refresh rural cultural development. By combining culture with art, artists' participation can bring new vitality to culture [54]. Ultimately, with the help of non-governmental organizations, culture and tourism will be integrated [34], and tourism can accelerate the protection and inheritance of culture.

China's rural revitalization became the national strategy in 2017. Culture restructuration is an important part even core of the strategy of rural revitalization. However, rural culture reconstruction is still facing big challenges based on our case. Particularly, rural culture gradually declined in most villages in China but a new culture has not matured, which is worrying. Rural culture always develops in many kinds of contradictions between the government and residents, between the space dominated by power and everyday life space, between recalling culture and exhibition propaganda. Even if residents tend to inherit cultural heritage, they often lack actions and supports. The excessive protection of ancient buildings has gradually lost residents' sense of place and identity as they often cannot access those buildings. They become relics and artefacts, rather than part of a living culture. Most residents (especially the elderly) are enthusiastic about the protection and inheritance of culture, yet the government subsidies are inadequate. Rural cultural reconstruction should focus on community, social coherence and local identity [32]. The actual cultural needs including everyday life space of residents have not received the government's attention and enough support. In the future, cultural reconstruction in the rural revitalization requires the residents' inheritance consciousness, the government's support, and the reasonable planning and protection as well as the engagement of non-governmental organizations.

References

1. Brennan MA, Flint CG, Luloff AE (2009) Bringing together local culture and rural development: findings from Ireland Pennsylvania and Alaska. Sociol Rural 49(1):97–112
2. Buser M (2012) The production of space in metropolitan regions: a Lefebvrian analysis of governance and spatial change. Plan Theory 11(3):279–298
3. Chen J (2007) Rapid urbanization in China: a real challenge to soil protection and food security. CATENA 69(1):1–15
4. Chen N (2016) Governing rural culture: agency, space and the re-production of ancestral temples in contemporary China. J Rural Stud 47:141–152
5. Chen M, Liu W, Lu D (2016) Challenges and the way forward in China's new type urbanization. Land Use Policy 55:334–339
6. Chen J, Wang Y, Wen J, Fang F, Song M (2016) The influences of aging population and economic growth on Chinese rural poverty. J Rural Stud 47:665–676
7. Chen R, Ye C, Cai Y, Xing X, Chen Q (2014) The impact of rural outmigration on land use transition in China: past, present and trend. Land Use Policy 40:101–110
8. Crouch D (1992) Popular culture and what we make of the rural, with a case study of village allotments. J Rural Stud 8(3):229–240
9. Deng X, Huang J, Rozelle S, Zhang J, Li Z (2015) Impact of urbanization on cultivated land changes in China. Land Use Policy 45:1–7
10. Friedmann J (2006) Four theses in the study of China's urbanization. Int J Urban Reg Res 30(2):440–451
11. Frisvoll S (2012) Power in the production of spaces transformed by rural tourism. J Rural Stud 28(4):447–457
12. Gorlach K, Klekotko M, Nowak P (2014) Culture and rural development: voices from Poland. Eastern Eur Countryside 20(1):5–26
13. Jenkins TN (2000) Putting postmodernity into practice: endogenous development and the role of traditional cultures in the rural development of marginal regions. Ecol Econ 34(3):301–314
14. Ju H, Zhang Z, Zhao X, Wang X, Wu W, Yi L, Wen Q, Liu F, Xu J, Hu S, Zuo L (2018) The changing patterns of cropland conversion to built-up land in China from 1987 to 2010. J Geog Sci 28(11):1595–1610
15. Kan K (2016) The transformation of the village collective in urbanising China: a historical institutional analysis. J Rural Stud 47:588–600

16. Kanbur R, Zhang X (1999) Which regional inequality? The evolution of rural–urban and inland–coastal inequality in China from 1983 to 1995. J Comp Econ 27(4):686–701
17. Kneafsey M, Ilbery B, Jenkins T (2001) Exploring the dimensions of culture economies in rural West Wales. Sociol Rural 41(3):296–310
18. Kung JKS (2002) Off-farm labor markets and the emergence of land rental markets in rural China. J Comp Econ 30(2):395–414
19. Lang W, Chen T, Li X (2016) A new style of urbanization in China: transformation of urban rural communities. Habitat Int 55(1):1–9
20. Lefebvre H (1991) The production of space (Trans: Smith N). Blackwell, Oxford
21. Lin J (1992) Rural reforms and agricultural growth in China. Am Econ Rev 82(1):34–51
22. Lin G, Xie X, Lv Z (2016) Taobao practices, everyday life and emerging hybrid rurality in contemporary China. J Rural Stud 47:514–523
23. Liu Y, Hu Z, Li Y (2014) Process and cause of urban-rural development transformation in the Bohai Rim Region China. J Geog Sci 24(6):1147–1160
24. Liu J, Liu Y, Yan M (2016) Spatial and temporal change in urban-rural land use transformation at village scale—a case study of Xuanhua district, North China. J Rural Stud 47:425–434
25. Li Y, Liu Y, Long H, Cui W (2014) Community-based rural residential land consolidation and allocation can help to revitalize hollowed villages in traditional agricultural areas of China: evidence from Dancheng county, Henan province. Land Use Policy 39:188–198
26. Li Y, Westlund H, Zheng X, Liu Y (2016) Bottom-up initiatives and revival in the face of rural decline: case studies from China and Sweden. J Rural Stud 47:506–513
27. Long H (2014) Land consolidation: an indispensable way of spatial restructuring in rural China. J Geog Sci 24(2):211–225
28. Long H, Heilig G, Li X, Zhang M (2007) Socio-economic development and land-use change: analysis of rural housing land transition in the Transect of the Yangtze River China. Land Use Policy 24(1):141–153
29. Long H, Li Y, Liu Y, Woods M, Zou J (2012) Accelerated restructuring in rural China fueled by "increasing versus decreasing balance" land-use policy for dealing with hollowed villages. Land Use Policy 29(1):11–22
30. Long H, Tu S, Ge D, Li T, Liu Y (2016) The allocation and management of critical resources in rural China under restructuring: problems and prospects. J Rural Stud 47:392–412
31. Long H, Zou J, Pykett J, Li Y (2011) Analysis of rural transformation development in China since the turn of the new millennium. Appl Geogr 31(3):1094–1105
32. Lysgard H (2016) The 'actually existing' cultural policy and culture-led strategies of rural places and small towns. J Rural Stud 44:1–11
33. McGee TG (2009) Interrogating the production of urban space in China and Vietnam under market socialism. Asia Pac Viewp 50(2):228–246
34. Molden O, Abrams J, Davis EJ, Moseley C (2017) Beyond localism: the micropolitics of local legitimacy in a community-based organization. J Rural Stud 50:60–69
35. O'Brien KJ, Li L (1999) Selective policy implementation in rural China. Comp Polit 31(2):167–186
36. Qian W, Wang D, Zheng L (2016) The impact of migration on agricultural restructuring: evidence from Jiangxi province in China. J Rural Stud 47:542–551
37. Rao J, Ye J (2016) From a virtuous cycle of rural-urban education to urban-oriented rural basic education in China: an explanation of the failure of China's rural school mapping adjustment policy. J Rural Stud 47:601–611
38. Ray C (1998) Culture, intellectual property and territorial rural development. Sociol Rural 38(1):3–20
39. Shanghai Minhang District Agricultural Committee (SMDAC) (2017) Construction and acceptance criteria of beautiful countryside in Minhang District. Retrieved from http://xxgk.shmh.gov.cn/mhxxgkweb/html/mh_xxgk/xxgk_ghj_ghjh_gl/2018-06-22/Detail_46821.htm
40. Shanghai Minhang District Local Chronicle Compilation Committee (SMDLCCC) (2011) Shanghai Minhang district chronicle 1992–2011: acceptance draft. People's Publishing House, Shanghai

41. Sicular T, Yue X, Gustafsson B, Li S (2007) The urban-rural income gap and inequality in China. Rev Income Wealth 53(1):93–126
42. Song W, Han Z, Deng X (2016) Changes in productivity, efficiency and technology of China's crop production under rural restructuring. J Rural Stud 47:563–576
43. Wang C, Huang B, Deng C, Wan Q, Zhang L, Fei Z, Li H (2016) Rural settlement restructuring based on analysis of the peasant household symbiotic system at village level: a case study of Fengsi Village in Chongqing, China. J Rural Stud 47:485–495
44. Wang Y, Yao Y (2003) Sources of China's economic growth 1952–1999: incorporating human capital accumulation. China Econ Rev 14(1):32–52
45. Woods M (2005) Rural geography: processes, responses and experiences in rural restructuring. Sage, London
46. Woods M (2010) Performing rurality and practising rural geography. Prog Hum Geogr 34(6):835–846
47. Woods M (2012) Rural geography III: rural futures and the future of rural geography. Prog Hum Geogr 36(1):125–134
48. Yang XJ (2013) China's rapid urbanization. Science 342(6156):310
49. Yang R, Xu Q, Long H (2016) Spatial distribution characteristics and optimized reconstruction analysis of China's rural settlements during the process of rapid urbanization. J Rural Stud 47:413–424
50. Ye C, Liu Z, Cai W, Chen R, Liu L, Cai Y (2019) Spatial production and governance of urban agglomeration in China 2000–2015: Yangtze River Delta as a case. Sustainability 11(5):1343
51. Ye C, Chen M, Chen R, Guo Z (2014) Multi-scalar separations: land use and production of space in Xianlin, a university town in Nanjing China. Habitat Int 42:264–272
52. Ye C, Chen M, Duan J, Yang D (2017) Uneven development, urbanization and production of space in the middle-scale region based on the case of Jiangsu province, China. Habitat Int 66:106–116
53. Ye C, Ma X, Cai Y, Gao F (2018) The countryside under multiple high-tension lines: a perspective on the rural construction of Heping village, Shanghai. J Rural Stud 62:53–61
54. Zacharias J, Lei Y (2016) Villages at the urban fringe–the social dynamics of Xiaozhou. J Rural Stud 47:650–656
55. Zhang Y, Li X, Song W, Zhai L (2016) Land abandonment under rural restructuring in China explained from a cost-benefit perspective. J Rural Stud 47:524–532
56. Zhang K, Song S (2011) Rural-urban migration and urbanization in China. China Econ Rev 14(4):386–400

Chapter 9
Marginalized Countryside in a Globalized City: Production of Rural Space of Wujing Township in Shanghai, China

Abstract This paper takes Wujing Township of Shanghai as a typical case to examine the process and dynamics of the production of rural space in China. Although Shanghai has generally become more urbanized and globalized than ever, the rural area in Shanghai is relatively marginalized. Production of space, as a social theory focusing on interactions between capital, power and class and their impacts on urbanization, is applied into one such micro-scale case of community development. There are three categories of social space separations in Wujing. The main driving force of producing these separations is power: top-down policymaking represses bottom-up community self-organizing. Rising housing prices driven by capital not only makes the locals only care for benefit from housing demolition, but also enlarges the gap between the locals and the migrants. It is the increasing strength of capital and power and weakened local voices that undermine community-based social space in Wujing.

9.1 Introduction

Rapid urbanization in recent years has changed traditional rural–urban relations greatly. Although levels of urbanization differ across developed and developing countries, the expansion of cities can be seen across the board. At the same time, rural restructuring and redevelopment have become large challenges for most countries—even in countries that are 80% urbanized—because the countryside is often overlooked [1, 29]. China has experienced rapid urbanization since its 1978 economic reform, leading to large changes in land use and population throughout the country [3, 21, 51]. China's rapid urbanization happened nationwide [53]. The comprehensive level of urbanization in China has seen a continuous increase in economic growth and a greatly changing geographical landscape [5, 51]. With such uneven development during this process, many environmental and social problems like the widening rural–urban gap have emerged [4, 8, 14, 19, 26]. The economic and social developments have reshaped even the rural areas, making rural redevelopment a key issue in

---

This chapter is based on [*International Development Planning Review*, Ye, C., Ma, X., Chen, R., & Cai, Y. (2019). Marginalised countryside in a globalised city: production of rural space of Wujing Township in Shanghai, China. *International Development Planning Review*, 41(3), 311–328].

© The Author(s), under exclusive license to Springer Nature Singapore Pte Ltd. 2023
C. Ye and L. Zhuang, *Urbanization and Production of Space*, Urban Sustainability,
https://doi.org/10.1007/978-981-99-1806-5_9

national policy [33, 52]. These shortcomings have resulted in some arguing that the "beautiful village planning" from 2005 focuses only on the aesthetic of rural areas, ultimately failing to support these communities [16]. From an institutional or political view, the rural–urban dualism of land ownership and household registration is the primary reason for resistance towards rural restructuring [25]. Rural development, then, is not only about the rural itself, but also related to the urban and urbanization, rendering present-day rural challenges more complicated than ever. Rural reconstruction, which is influenced by national economic and national policies as well as globalization, needs a social theory to analyze [23].

Rural areas are heterogeneous, dynamic and dystopic [6]. These key characteristics make rural research a cross-disciplinary field of study. Rural research should mainly focus on rurality, rural space and rural identities connecting multi-disciplinary theories and methods including geography, sociology, economics, planning and so on [27, 34, 45, 46]. In the past, rural research has often been carried out on a single perspective like politics, economics, sociology and geography [25, 44, 48], but lack of an integrated rural theory. Therefore, production of space, which integrates the theories of political economics, human geography and sociology into urbanization research, is a suitable theory to apply to rural research. Production of space mainly means capital, power and class shape urbanization and finally make urban space and urbanization become their product, which becomes an important theory to study the urban issues and urbanization since 1970s [12, 13, 38, 39]. Production of space is widely used for some empirical cases on urbanization [2, 17, 30–32, 43, 50]. However, contrary to rich theoretical and practical achievements on the urban, the study on production of rural space is still extremely limited, except for several valuable explorations [9, 11]. Marsden [28] pointed out that rural geography needs to be rethought from the angle of the interaction between space and society, between the natural and the social. In fact, as a link between the rural and the urban, production of rural space has a natural and necessary connection with urbanization or the urban space. Thus, the bias ignoring the countryside should be revised and reversed [42], especially research about the relations between urbanization and production of space.

Shanghai is a globalized city that has undergone a diversified urban–rural transition in the process of urbanization since 1990s. Contrary to common belief, rural areas in Shanghai are still lagging behind the city center. Globalization and urbanization help improve Shanghai's economic growth and a large number of migrants, at the same time, brings people under pressure of high housing prices, which produces more change and uncertainty for the local community. Even the villages of Wujing, have been influenced by them. In this article, we also describe Shanghai's influence on Wujing, especially from high housing prices and migrants. Wujing is a typical case of a marginalized town with five villages in Shanghai, and this article seeks to apply the production of space theory into practice to explore this marginalization process, and the forces and dynamics of production that are acting on this rural space. In the article, production of space in Wujing takes place together with new rural planning. Based on the period of the planning "beautiful countryside construction of Wujing", it began in December 2016, and ended in early 2018.

Table 9.1 Valid questionnaire information

Item	Category and proportion
Sex ratio	Male: 52% Female: 48%
Age distribution	13% (<25) 26% (26–40) 35% (41–55) 26% (>56)
Type of resident	Local: 57% Nonlocal: 43%

Source Authors' survey

9.2 Methodology

9.2.1 Research Method

This paper draws its data from statistical yearbooks and a questionnaire. Minhang Statistical Yearbook [36] and Shanghai Statistical Yearbook [35] are used to grasp the basic situation in Wujing and serve as the research background and analysis basis. A simple survey of residents' daily lives was conducted. A total of one hundred questionnaires was issued, and ninety-two valid questionnaires were collected. The details of the valid samples are shown in Table 9.1. In terms of qualitative research, the authors used interviews to understand the residents' attitudes towards community construction, the residents' living conditions, neighborhood relations and so on. At the same time, the authors participated in the project named "Beautiful Countryside Construction of Wujing", in order to understand rural construction from the government's perspective. In short, through these materials and methods, the authors can analyze the production of space in Wujing from various aspects.

9.2.2 A Framework on Urbanization and Production of Rural Space

Before the 1970s, space is often seen as a physical existence or like a kind of container without value judgments, which ignores the individual, political and social relations shaping "space" and so failing to provide reasonable explanations for the interactions between the social and space [37, 49]. With the core argument that "(social) space is the (social) product", production of space was put forward in the 1970s by the French Marxist philosopher and sociologist Henri Lefebvre and the production of space is deemed as one of the most remarkable spatial theories [18]. Space and social relations are linked together: the different classes or societies produce their spaces; at the same time, these spaces are also shaping or influencing social relations and process.

The theory of production of space means the urban space is constantly reshaped by the political, economic and social factors which are mainly embodied by power, capital and class [50]. Power acts through policies, discourses or governance. Capital runs through distributing resources and dominates uneven development in different regions. The process of urbanization includes high-income and low-income groups and these groups occupy different spaces. In other words, different social groups produce different social spaces so that the gap among the social groups widens. Shanghai is a typical case to reflect on urbanization and production of space. In particular, the rural development of Shanghai also reflects the process of production of space.

This article attempts to analyze the production of space of Wujing and explore the factors that shape rural development. Urbanization is mainly represented with a transforming process from rural space to urban space. It should be noted that Wujing is influenced by the development of Shanghai, thus the production of space in Wujing should be analyzed in the context of globalization and urbanization. The core meaning of production of space lies in changing social relations produced by power and capital as the dominant forces that influence and divide social classes. Hence, the space reflects not only the scene and location of things but also the products of a set of specific social relations interaction. The production of rural space is dynamic and a simple framework (Fig. 9.1) can be used to explain the relationship between urbanization and production of space. Urbanization and production of space are regarded as two systems which interact and intertwine. The relationship between them is shaped and reshaped by each other.

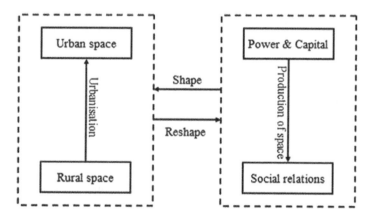

Fig. 9.1 The framework on the relation between production of space and urbanization

9.3 Production of Space in Wujing

9.3.1 Wujing Town as a Case Study

The rural construction of Wujing is a typical case of new rural construction. First, Wujing is in Minhang District of Shanghai, which is one of the most developed areas in China, and also has a rich cultural heritage. Shanghai is an important economic, transportation, science and technology hub in China. Shanghai cooperates with Jiangsu, Zhejiang and Anhui to form the Yangtze River Delta of China urban agglomeration. However, as a global city, Shanghai's villages are not developed well.

Secondly, Wujing contains towns that have been transformed by urbanization and industrialization, as well as villages in cities and pure countryside. For this reason, studying this area will involve different rural constructions and provide more valuable materials. Table 9.2 shows the comparison between Minhang District—the administrative unit of Wujing—and Shanghai in 2015.

Thirdly, Wujing is a town near to the center of the city and contains several institutions (Fig. 9.2). A variety of cultural and scientific institutions are located in Wujing, including universities, science parks and a national high-tech industry developmental zone. Accordingly, Wujing is a diverse and complex area. It is actually both a rural space and an industrial base. Moreover, compared with the developed inner city without countryside, Wujing is located at the suburb and has five villages, which is often called the "countryside" of Shanghai by local people. This also means Wujing is marginalized in Shanghai. Meanwhile, Wujing is suffering from the impact of migrants, bringing much change and uncertainty to the local community. According to the interviews, most of the migrants come from Anhui Province and mainly work

Table 9.2 Comparison between Shanghai and Minhang District in 2015

Index	Shanghai	Minhang District
Permanent population (ten thousand)	2415	254
Household population (ten thousand)	1443	107
Migration (person)	62,789	9548
Per capita disposable income (yuan)	49,867	50,912
GDP (100 million yuan)	25,123	1965
Primary industry GDP	110	1
Second industry GDP	7991	1014
Third industry GDP	17,023	950
Grain output (10,000 tons)	112	1
General public budget revenue (100 million yuan)	5520	628
General public budget expenditure (100 million yuan)	6192	344
Number of downstream units	17	14

Data source SBS [35], SMDBS [36]

in nearby factories or are working as waiters. Because of high housing prices, they have to rent, which brings income to the local villagers. There seems to be a contradiction between locals and migrants. The migrants complain that the locals always ignore them; however, the village committee also report that migrants often do not cooperate with management. Hence, the new rural construction of Wujing involves many important factors and aspects.

The villages in Wujing Township have both similarities and differences with other traditional Chinese villages. On the one hand, like other Chinese rural areas, the villages in Wujing have a village committee. Grass-roots organizations are very important in China and are established by democratic elections, which can protect the rights of rural residents and realize the autonomy of villagers. The main tasks of these committees are to mediate residents' disputes, to maintain social order and to convey the villagers' opinions, demands and suggestions to the town government. Despite such roles, these village committees are also under the jurisdiction of the town government. On the other hand, due to the impact of urbanization and globalization

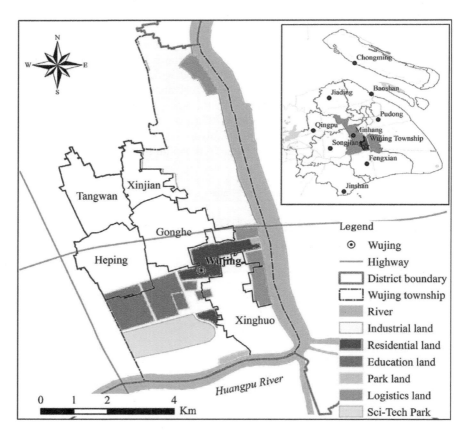

Fig. 9.2 Location of Wujing township in Shanghai. *Source* SMDBS [36]

9.3 Production of Space in Wujing

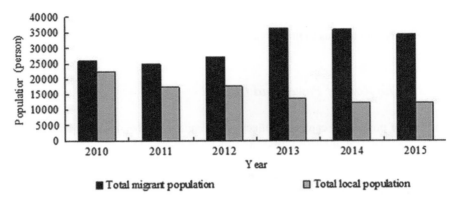

Fig. 9.3 Migrants and the local people of Wujing from 2010 to 2015. *Source* SMDBS [36]

in the city, these villages are also unique. Shanghai, being one of the largest cities of China attracts many migrants. Thus, Wujing shows a widespread phenomenon of a larger migrant population than local population (Fig. 9.3).

The rural development of Wujing has been affected by various policies and experienced a lot of "new rural construction", referring to economic, political, cultural and social construction of the countryside. At the same time, Wujing is striving for "beautiful countryside construction" and the "establishment of a national civilized city"; they are managed by different departments. In view of these aspects, the rural development of Wujing is influenced by various social classes, which are constantly affected by the power and capital. The process and dynamics of production of rural space construction need to be further explored.

9.3.2 The Characteristics and Dynamics of Spatial Production in Wujing

The core issue of rural development in China is social space, and the basic unit in China is the community. Social space is perceived and used by social groups, which can reflect social values, preferences and pursuit [15]. However, social space is also the product of external forces. Power and capital play an important role in promoting the development of rural community in China and shaping the social space. The rural population is regarded as the main source of labor for the urban areas. Since 2005, the Chinese government has implemented numerous policies to build the new socialist countryside with the goal of "advanced production, improved livelihood, civilized social atmosphere, cleaning and tidy villages and efficient management", which emphasizes "people-oriented" and "local conditions" in the construction process [24]. However, in actuality, "people-oriented" and "local conditions" are often overlooked or cannot be put into practice effectively. The reason for this is closely related to the role of power and capital in the process of urbanization. Wujing Township, as a case study, can help us to find a reasonable explanation.

9.3.3 The Role of Power and Capital

The role of power and capital is reflected in shaping the social space of Wujing. In China, the manifestations of power are official policies and institutions that cannot be violated; meanwhile, different classes in fact hold different powers. For example, the rural construction in Wujing is under the jurisdiction of the Minhang District Government, the Wujing Town Government and the village committee. Capital refers to "an asset to be mobilized by a group, individual or institution as wealth … it is not a thing, but is a "social relation" that appears in the form of things (money, means of production)" [10]. For example, locals have the capital of real estate compared with migrants. In Wujing, the flow of capital among different classes is not balanced either. Actually, in rural–urban planning, the governments of different levels including the village committee dominate and decide who can make the planning decisions, who can be included in the planning process and which village can get more funds. The combination of power and capital shaped the social space of Wujing.

Separations exist in productive spaces, living spaces and ecological spaces. Many villages of Wujing have rivers passing through, which is a very important natural space. However, the residents of Wujing are unable to gain direct access to the river as the river is cordoned off for security considerations. In the planning process, we learned that these rivers are designed for government to achieve the new rural construction objectives of beautifying the township. As a consequence, this measure does not actually benefit the residents significantly. Similarly, many woods in the town were also surrounded by steel wires, resulting in very clear separations between the roads, rivers, houses, forests and fields in the villages (Fig. 9.4). The benefits of natural spaces to the livability of the township are thus limited due to the lack of access.

Fig. 9.4 Division of the road, river and forest land. *Source* Authors' survey

9.3 Production of Space in Wujing

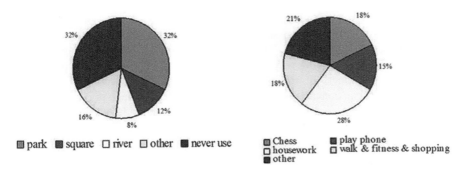

Fig. 9.5 Frequency of use of the public amenities and types of residents' daily activities. *Data source* Authors' survey

In addition, a survey was conducted to investigate the types of residents' daily activities and the frequency of use of the public amenities (Fig. 9.5). Housework accounted for a considerable proportion in the daily activities' survey amongst married and middle-aged women: middle-aged women spend their time taking care of children, cooking and other household chores, whereas younger married women take care of the children when the grandparents are not available to help, and spend their time working and doing housework. Chess, walking and fitness activities are other popular leisure activities of residents. The survey shows that most people use the park frequently corresponding to the residents' preferred leisure activities. In addition, some young people do not have time to go to these public spaces because of their busy work. In sum, most of the residents have a very strong demand for public activity spaces. If the community does not provide certain public places for residents to use, then the residents who need the activity spaces will have a lower overlap. The separation of space will lead to social space problems [41].

The universities, scientific park, urban village and villages of Wujing are separated. Wujing is home to many world-class tertiary institutions such as Shanghai Jiao Tong University, East China Normal University and the scientific park of Wujing. These institutions give Wujing rich scientific and technological resources. However, the universities and the high-tech Park in Wujing do little work on village development or for residents. For example, Heping village, which is located north of the universities and the scientific park only 3–7 km away, did not get the help from them in the construction of the village. Also, the universities and the scientific park have taken measures to ensure the quiet and safety inside of their locations, such as building walls and erecting "no entry" signs for foreign vehicles. These measures lead to the separation between city roads and private roads inside. Pedestrians and cars can only bypass the line; these measures also caused the separation of internal resources and external accessibility. The spatial isolation between these institutions and the town result in the lack of communication and interaction between various social groups. In addition, in the "Minhang 2035 Plan", Wujing Town is positioned as the core area of scientific and technological innovation, which indicates that the plan will promote the development of science parks, scientific research institutions and universities. Wujing's rural space is being marginalized, uneven and perhaps even disappearing in the future. These helps explain why Heping village has not developed well despite being very close to the universities and scientific park.

One of the products of rapid urbanization is the "urban village", which is surrounded by urban construction land. Despite the urban village playing a positive role and undertaking some responsibilities previously shouldered by city governments, they are still viewed as an unregulated asset [20]. There is an increasing migration trend of wealthier villagers moving to urban areas due to the good medical and educational levels and more social security. Meanwhile, the rural areas, with its low-cost housing, attracted some migrants who came to Shanghai from other provinces to live here. Hence, the social structure of the countryside consists mainly of the elderly, the poor and the migrants. In the process of urban village renovation, universities did not fully consider the equal supply of education; the scientific park has not provided jobs for rural people. That is why the rural areas cannot benefit from the universities and the scientific park, which is mainly due to government policies. In their policies, the government favored the development of enterprise to increase local economic income and government taxation, but without considering how this would benefit the residents in areas such as employment and social services.

There is a large estrangement between migrants and locals. In 2015, there were 12,151 locals and 34,394 migrants in Wujing, and migrants accounts for 74% of the total population [36]. The migrants and local people lack communication, which can be largely attributed to the migrants being too busy with work because of needing to earn money to pay rent. This leaves little time for locals and migrants to interact, let alone build a harmonious community. The lack of effective communication results in considerable misunderstandings between the locals and migrants, and a lack of common meeting space further exacerbates the issue. The household registration system is an important rigid constraint, meaning migrants cannot get the local household registration accounts they need and so weakening their own local identity, which may cause more conflicts and contradictions.

The production of rural space is influenced by power. Wujing was a famous industrial base and many chemical companies are located here. However, with the development of Minhang Riverside and transformation of the old industrial park, most companies have been ordered to stop or to be relocated. Several remaining enterprises, such as thermal power plants, have not yet moved away. The reason is that the relocation is a gradual process. What is more, they are large state-owned enterprises, which play an important role in regulating the national economic goals, thus, their shutting down and relocation involves a wide range of interests.

Capital is often distributed unequally by power. The housing prices in Wujing Township have been on an upward trend in recent years (Fig. 9.6). The price increase affects the residents in different ways. The migrants who rent houses must spend more time working to make money for the rising rent. Therefore, the migrants' sense of belonging is weak, making them not care about how to rebuild the village. The locals who have their own property hope the house prices continue to rise and then be demolished to get good compensation or better housing, and so they do not care about

9.3 Production of Space in Wujing

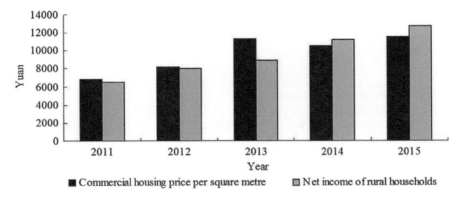

Fig. 9.6 The trend of commercial housing price and per capita net income of rural households in Wujing. *Data source* SMDBS [36]

the construction of new socialist countryside either. Since almost all residents, then, are not concerned about the countryside construction, the new countryside building seems to make no sense in the end.

From the rural construction in Wujing, we can see that urbanization has promoted rural development. The policy of rural development has accelerated the changes in rural areas and improved the living environment in the countryside, which is at least cleaner and tidier than before. Different classes have different power and capital in this process, which makes them have different views on rural construction, thus affecting the process and effect of rural construction. As a result, the social space in Wujing is the result of the interaction of power and capital. Furthermore, the villagers' committee does not represent villagers' opinions effectively. The committee is not only prejudiced against migrants but must also accept decisions from the town government, which has overall jurisdiction. As a result, the committees do not convey the needs of residents to the powers above and just to achieve some policy objectives. The role of power is largely overwhelming and reflects across all aspects of the space, housing, life, work, etc. Capital has further deepened the separation situation that has already been formed. Eventually, the social classes are separated in the production of space.

9.3.4 The Marginalized Villages

The reasons leading to marginalized rural areas have received little attention. Table 9.3 shows that the countryside has a much lower residents' income and less public infrastructure than the urban areas. It is noted that Minhang is the suburb of Shanghai, so the more distant suburbs could be even worse. Capital and power were only concentrated in a few places, whilst the places which need to develop did not receive government attention and financial support. The concentration of power and capital

Table 9.3 Minhang District rural residents have lower income and public facilities

	2014		2015	
	Minhang District	Shanghai	Minhang District	Shanghai
Per capita disposable income of rural households (Minhang District)/urban households (Shanghai)	27,560	47,710	30,130	52,962
Number of medical institutions	27	4987	28	5016
Number of schools	317	1724	319	1748
Number of libraries	1	25	1	25
Number of fitness facilities	1344	11,091	1360	9905

Data source SBS [35], SMDBS [36]

has made new countryside construction deviate from the "people-oriented" goal ("people-orientated" means "take human development or human daily life as the most fundamental thing").

At the same time, under the domination of power and capital, the community lacks vitality and cohesion. The new rural construction only concerns roads, environment and some other aspects of the villages' appearance. Other visible aspects, such as whether the village's living environment is clean and whether there is garbage are important indicators. Further, the implementation process has different standards in different villages according to the basic conditions of the villages. The constructions of the community and public spaces are neglected. Consequently, people—who are the key elements in the community—are marginalized while the external environment and the aesthetics of beauty take central position. However, it is worth noting that if "people-oriented" cannot be implemented to a specific person and each type of person, then this statement will become ineffective.

Due to the construction of new socialist countryside and the role of high housing prices, locals are interested in "demolition" because they might get better housing or higher compensation. The migrants are concerned about the housing rent, welfare and social spaces for their children and parents, but in fact, there are few public spaces in the community. In this community, both locals and migrants have a low sense of belonging and so the community is marginalized.

9.4 Discussion

The construction of new socialist countryside in China often emphasizes "people-orientation" and "development based on local conditions", which are actually contradictory in practice due to the force of power and capital. "People-oriented" emphasizes the interests and needs of residents, which include both locals and migrants. However, in the reality of rural planning and development, the governments' power plays a pivotal role. Government has to achieve construction indicators rather than

9.4 Discussion

prioritize the interests of residents, so the residents have a very low sense of belonging. Coupled with the rising capital-driven real estate market prices, most of the locals are only concerned about demolition, while the migrants are more concerned about rising rents. "Local conditions" emphasizes that the development should be based on the specific local base and context. However, the local village of Wujing has gradually disappeared, and the new local characteristics have not been established. The construction of new socialist countryside neither achieves the aim of "people-oriented" nor the "local conditions". Most importantly, in fact, the pattern of production of space is contradictory to "people-oriented" and "development based on local conditions".

The combined effect of the power from the government and capital mainly represented by high housing prices makes the rural construction only superficial, rather than help promote community integration and any realization of the notion of "people-oriented". The community building is not based on public opinion and is not developed in a bottom-up way. On the contrary, the pattern of top-down administrative power is more popular and dominant than others, and is the biggest problem in rural reconstruction in China. As [22] indicated that the rural development policy was in essence a top-down strategy based on state intervention and it may be encountered resistance in the implementation process, however, the rural development policy would protect farmers' rights and mitigate against rural protests when incorporating elements of "bottom-up planning" into the strategy.

New rural (re)construction should initially be based on local conditions and should implement the different policies. However, the top-down pattern of uniform standardization does not fully consider local needs, instead only satisfying the so-called official standard. In the process of rebuilding countryside, the government always has the greater power, including discourse, distribution and planning, so the top-down policies play a decisive role. In addition, due to agriculture being more dispersed and agricultural income being lower than in urban areas, the residents do not care for pursuing the diversified life they want, and lose the sense of belonging. Under the pressure of rising housing prices driven by capital, the locals and migrants have no care or time for community. The gap between migrants and locals is enlarging so that building a harmonious community becomes a meaningless discussion in the end. The rural landscape has become the mirror of the discourse on government power.

From the case of Wujing, we can draw inspiration and guidance to achieve "people-oriented" in the process of rural construction. The production of space is mainly manifested by the joint function of power and capital, so we will start there. First of all, power exists in government, village committees, social organizations and with residents. By raising the voice of social organizations and residents, it can balance the power away from the government. For example, the government could set up a policy that village construction must solicit opinions from residents, and select representatives of different classes to participate in the discussion on the rural construction. Secondly, the government should balance the capital allocation between locals and migrants. For the locals, a reasonable compensation policy for house demolition can be set up, for example people whose houses have been demolished could only get a compensatory house in their previous residence, rather than in the city center. For the migrants, the government should provide a subsidy policy for renting and provide equal rights for education, medical treatment and so on. Finally, the community should strengthen the construction of public space and organize more community activities. Incentives should be used to motivate more residents to participate in community activities, so as to increase the migrants' sense of belonging and promote the harmony of the community.

9.5 Conclusions

With the progress of urbanization, new rural problems in the world are constantly emerging and evolving. We are facing a challenge to rebuild countryside and for rural redevelopment in the future. The rural is a complex synthesis and must be considered from various viewpoints. Rural reconstruction should be done from a cross-disciplinary and time–space perspective. Cross-disciplinary perspective can help us make the connection with the understanding of rural history and rural future [47].

Production of space is a multi-disciplinary and appropriate theory from which to understand rural development in the process of urbanization, which provides us with a new perspective and methodology to re-examine rural issues. There is a contradiction between the governmental goals of "people oriented" rural development and the ways in which rural space is actually being transformed. Although the latter is emphasized by the main ideology, the pattern of production of space in China always prevails and destroys the people-oriented base. According to the Wujing Town case study in Shanghai, we have gained new insights into the relationship between urbanization, rural community and production of space in China. Power and capital are two key forces or factors playing a dominant role in the production of space. The top-down policies are the initial force of rural development and the governmental officials hold more power than the residents. Some basic principles such as "people-oriented" and "development based on local conditions" are often ignored in the new rural construction. There is no consensus on community building.

The top-down tradition of countryside planning is changing, and a few residents also have been included in the plan-making process. However, the government-dominated pattern is not changed, and a great deal of rural construction funds come from government, which has weakened the interests of residents in planning. The reconstruction of villages and planning of the rural community should link to all stakeholders, upgrading the strengths of the locals, migrants and local communities in order to balance out the forces from power and capital. In the end, to achieve sustainable urbanization is the most important thing in the future [40]. Production of space is not only theoretical but also practical, and tries to intervene in or change society [7]. This reminds us of the importance of methodology. In fact, the scholars, planners and researchers are residents, practitioners and citizens in the urban or the rural too. Thus, the research will become more complex than ever, however, it probably is the proper beginning to approach the real world.

References

1. Antrop M (2004) Landscape change and the urbanization process in Europe. Landsc Urban Plan 67(1–4):9–26
2. Buser M (2012) The production of space in metropolitan regions: a Lefebvrian analysis of governance and spatial change. Plan Theory 11(3):279–298
3. Chen G, Glasmeier AK, Zhang M, Shao Y (2016) Urbanization and income inequality in post-reform China: a causal analysis based on time series data. PLoS ONE 11(7):e0158826
4. Chen M, Huang Y, Tang Z, Lu D, Liu H (2014) The provincial pattern of the relationship between urbanization and economic development in China. J Geog Sci 24(1):33–45
5. Chen M, Lu D, Zha L (2010) The comprehensive evaluation of China's urbanization and effects on resources and environment. J Geog Sci 20(1):17–30
6. Cloke P (2003) Country visions. Prentice Hall, London
7. Cresswell T (2013) Geographic thought, a critical introduction. Wiley-Blackwell, Chichester
8. Cui L, Shi J (2012) Urbanization and its environmental effects in Shanghai, China. Urban Climate 2:1–15
9. Frisvoll S (2012) Power in the production of spaces transformed by rural tourism. J Rural Stud 28(4):447–457
10. Gregory D, Johnson R, Pratt G, Watts M, Whatmore S (2009) The dictionary of human geography, 5th edn. Blackwell, Oxford
11. Halfacree K (2007) Trial by space for a "radical rural": introducing alternative localities, representations and lives. J Rural Stud 23(2):125–141
12. Harvey D (1973) Social justice and the city. Edward Arnold, London
13. Harvey D (1985) The urbanization of capital. Blackwell, Oxford
14. Hua L, Ma Z, Guo W (2008) The impact of urbanization on air temperature across China. Theoret Appl Climatol 93(3–4):179–194
15. Johnston R, Gregory D, Smith D (1994) The dictionary of human geography, 3rd edn. Blackwell, Oxford
16. Lang W, Chen T, Li X (2016) A new style of urbanization in China: transformation of urban rural communities. Habitat Int 55:1–9
17. Leary M (2013) A Lefebvrian analysis of the production of glorious, gruesome public space in Manchester. Prog Plan 85:1–52
18. Lefebvre H (1991) The production of space (Trans: Smith N). Blackwell, Oxford
19. Li Y, Long H, Liu Y (2015) Spatio-temporal pattern of China's rural development: a rurality index perspective. J Rural Stud 38:12–26
20. Liu Y, He S, Wu F, Webster C (2010) Urban villages under China's rapid urbanization: unregulated assets and transitional neighbourhoods. Habitat Int 34(2):135–144
21. Liu Y, Yin G, Ma L (2012) Local state and administrative urbanization in post-reform China: a case study of Hebi city, Henan province. Cities 29(2):107–117
22. Long H, Li Y, Liu Y, Woods M, Zou J (2012) Accelerated restructuring in rural China fueled by "increasing versus decreasing balance" land-use policy for dealing with hollowed villages. Land Use Policy 29(1):11–22
23. Long H, Liu Y (2016) Rural restructuring in China. J Rural Stud 47:387–391
24. Long H, Liu Y, Li X, Chen Y (2010) Building new countryside in China: a geographical perspective. Land Use Policy 27(2):457–470
25. Long H, Tu S, Ge D, Li T, Liu Y (2016) The allocation and management of critical resources in rural China under restructuring: problems and prospects. J Rural Stud 47:392–412
26. Long H, Zou J, Pykett J, Li Y (2011) Analysis of rural transformation development in China since the turn of the new millennium. Appl Geogr 31(3):1094–1105
27. Marsden T (1994) Opening the boundaries of the rural experience: progressing critical tensions. Prog Hum Geogr 18(4):523–531
28. Marsden T (1996) Rural geography trend report: the social and political bases of rural restructuring. Prog Hum Geogr 20(2):246–258

29. McCarthy J (2008) Rural geography: globalizing the countryside. Prog Hum Geogr 32(1):129–137
30. McGee TG (2009) Interrogating the production of urban space in China and Vietnam under market socialism. Asia Pac Viewp 50(2):228–246
31. Nasongkhla S, Sintusingha S (2013) Social production of space in Johor Bahru. Urban Stud 50(9):1836–1853
32. Nguyen M, Locke C (2014) Rural-urban migration in Vietnam and China: gendered householding, production of space and the state. J Peasant Stud 41(5):855–876
33. Rigg J (1998) Rural-urban interactions, agriculture and wealth: a Southeast Asian perspective. Prog Hum Geogr 22(4):497–522
34. Roche M (2003) Rural geography: a stock tally of 2002. Prog Hum Geogr 27(6):779–786
35. Shanghai Bureau of Statistics (SBS) (2016) Shanghai statistical yearbook. China Statistics Press, Beijing [上海市统计局. 2016. 上海统计年鉴. 北京: 中国统计出版社.]
36. Shanghai Minhang District Bureau of Statistics (SMDBS) (2011–2016) The Minhang statistical yearbook. China Statistics Press, Beijing [上海市闵行区统计局. 2011–2016. 闵行统计年鉴. 北京: 中国统计出版社.]
37. Smith N (1984) Uneven development: nature, capital, and the production of space. Blackwell, Oxford
38. Soja EW (1980) The socio-spatial dialectic. Ann Assoc Am Geogr 70(2):207–225
39. Soja EW (1989) Postmodern geographies: the reassertion of space in critical social theory. Verso, London
40. Tan Y, Xu H, Zhang X (2016) Sustainable urbanization in China: a comprehensive literature review. Cities 55:82–93
41. Wang D, Li F, Chai Y (2012) Activity spaces and sociospatial segregation in Beijing. Urban Geogr 33(2):256–277
42. Whatmore S (1993) Sustainable rural geographies? Prog Hum Geogr 17(4):538–547
43. Wilson J (2013) The urbanization of the countryside depoliticization and the production of space in Chiapas. Lat Am Perspect 40(2):218–236
44. Woods M (2008) Social movements and rural politics. J Rural Stud 24(2):129–137
45. Woods M (2009) Rural geography: blurring boundaries and making connections. Prog Hum Geogr 33(6):849–858
46. Woods M (2010) Performing rurality and practising rural geography. Prog Hum Geogr 34(6):835–846
47. Woods M (2012) Rural geography III: rural futures and the future of rural geography. Prog Hum Geogr 36(1):125–134
48. Yang R, Xu Q, Long H (2016) Spatial distribution characteristics and optimized reconstruction analysis of China's rural settlements during the process of rapid urbanization. J Rural Stud 47:413–424
49. Ye C, Chai Y, Zhang X (2011) The theory and research progress of space production and its implications for urban studies in China. Econ Geograph 31(3):409–413. [叶超, 柴彦威, 张小林. 2011. "空间的生产"理论、研究进展及其对中国城市研究的启示. 经济地理, 31(3): 409–413.]
50. Ye C, Chen M, Chen R, Guo Z (2014) Multi-scalar separations: land use and production of space in Xianlin, a university town in Nanjing, China. Habitat Int 42:264–272
51. Zhang K, Song S (2003) Rural-urban migration and urbanization in China: evidence from time-series and cross-section analyses. China Econ Rev 14(4):386–400
52. Zhang L (2008) Conceptualizing China's urbanization under reforms. Habitat Int 32(4):452–470
53. Zhou Y (2006) Thoughts on the speed of China's urbanization. City Plann Rev (S1):32–35+40. [周一星. 2006. 关于中国城镇化速度的思考. 城市规划, (S1): 32–35+40.]

Chapter 10
The Countryside Under Multiple High-Tension Lines: A Perspective on the Rural Construction of Heping Village, Shanghai

Abstract Different rural developmental options exist because of different national and regional contexts. Since its reform and opening-up policy, China has undergone rapid urbanization and development, and rural–urban relationships have become disjointed. China's attempts to change these circumstances are inadequate. A new movement known as "rural construction," which aims to promote or direct the social or economic development of rural areas, has been gaining momentum in recent years. This paper takes Heping village in Shanghai as a typical case of rural construction, and integrates the methods of statistics, field investigation, and personal participation into rural research. The village presents challenges related to multiple "high-tension lines," literally and figuratively. In a literal sense, physical high-tension lines are located near residents' houses; in a figurative sense, high-tension lines come from power and capital, symbolizing the predominant top-down mode of rural construction and imbalance of allocated funds in the process of rural planning; and finally, psychological high-tension lines make the residents lose their sense of belonging. The local governments often regard village construction projects as opportunities to showcase their work, failing to involve villagers in the decision-making process; movements such as beautification of the countryside lack bottom-up initiatives. However, a collaborative approach between the government and residents is ideal in rural planning and construction. The theories of rural research should be renewed under the context of the changing rural world. It is worth exploring some new methods like metaphor into rural studies.

10.1 Introduction

Urbanization in China has been a dominant focus of international attention [49], and has been studied by many scholars in different fields [3, 4, 7, 16, 20]. Urbanization brings about not only economic development but also different ecological

---

This chapter is based on [*Journal of Rural Studies*, Ye, C., Ma, X., Cai, Y., & Gao, F. (2018). The countryside under multiple high-tension lines: a perspective on the rural construction of Heping Village, Shanghai. *Journal of Rural Studies*, 62, 53–61].

© The Author(s), under exclusive license to Springer Nature Singapore Pte Ltd. 2023
C. Ye and L. Zhuang, *Urbanization and Production of Space*, Urban Sustainability,
https://doi.org/10.1007/978-981-99-1806-5_10

and social problems [36, 57, 58]. China's rapid urbanization since 1978 has significantly influenced its rural development, aggravating the conflict between the two categories of areas [2, 55]. The coordinated development of urban and rural areas in China declined significantly from 2000 to 2008, in certain developed areas, urban–rural development is particularly disjointed [14, 18]. China has carried out a series of "new rural construction" programs since 2005, titled diversely as "livable rural planning," "mountain support planning," and "beautiful countryside construction," all allegedly aiming to improve residents' lives, foster a civilized social atmosphere, and develop appealing and organized villages that can be efficiently managed [17]. However, these plans have ultimately failed to enhance rural areas as promoted.

Contemporary research of rural areas and geography has begun including the role of rurality, the future of rural space, and global issues such as climate change and food security [40, 45, 47, 48]. Since the 1990s, rural construction has become a popular issue among scholars [22, 23, 28, 38, 39, 42, 43]. Woods [46] defined rural reconstruction as an interconnected process of reshaping rural society, mobility, and economic structures. The process of rural construction in China is characterized by dynamic, multi-scalar, and hybrid thought [19]. In the context of urbanization, agriculture is undergoing a transformation precipitated by changes in industrialization and migration [29, 30, 37]. Driven by industrialization and urbanization, rural land use has been extensively restructured [15, 21, 54, 56].

Rural space is the essential issue in rural construction. A critical theory of urbanization is the production of space, widely considered in many urban studies [1, 5, 24, 27, 53]. While compared to the relative plethora of research concerning space in cities, there are few studies contemplating rural production of space [8, 9], although some scholars have explored the theory of production of space as a tool to interpret the space and politics of the countryside [41, 51].

The construction of urban and rural space in China is accelerating. According to the 2016 Statistics Bulletin on Urban–Rural Construction, there were 20,883 towns in China, with 958 million people registered as village residents; China's investment in village construction occupies first place in its total investment in the construction of villages and towns [25]. China has paid increasing attention to the development of the countryside, with moderate success, but some villages in China remain in crisis. As a global megalopolis, Shanghai has experienced rapid urbanization and globalization; however, rural development in Shanghai is affected by various factors such as new policies, large numbers of migrants, and access to funds. Thus, Shanghai's circumstances can be seen as a kind of hybrid of urban and rural development.

"Representations of rurality remain a significant feature in the spatialization of everyday discourse, and therefore remain a legitimate focus for investigation in rural studies" [6]. In Shanghai, development of the rural area is not met with optimism. In particular, rural life exists under the "multiple high-tension lines" situation: there are the physical high-tension lines, as well as the high-tension lines in the power, capital, and psychological senses, decreasing the residents' sense of belonging, and causing a general lack of interest that hinders Shanghai's rural construction. Therefore, taking Heping Village of Wujing Town in Shanghai as an example, this paper aims to uncover and specify the realities of China's rural redevelopment.

Rural construction is a complex process that involves many aspects of social and economic activities. Previous studies on rural construction often elaborated upon the rural elements, such as systems and classes, and failed to consider any theory integrating social-economic factors including offering insufficient critiques. Therefore, based on the critical theory of production of space which is defined as the interaction between the urban and its space changed by capital or power [52], this paper aims to answer the following questions: "What is the current status of rural construction in China?", "What factors led to such a result?", and finally, "What are the focuses of China's rural construction in the future?" In addition, this paper provides two new perspectives. Compared with previous statements on rural issues, especially regarding China, we analyze rural construction in China from a critical perspective. More importantly, we use a metaphor, the high-tension line, to narrate and explain rural construction in Shanghai. In this paper, the high-tension line serves as fact, landscape, and metaphor. On the one hand, the physical high-tension line exists in reality. On the other hand, it symbolizes the high-tension line created by an imbalance of power and capital, as well as that in an individual's spirit. Metaphors such as this are rarely used in current rural studies except as by Cloke [6], whose Country Visions' cover in particular illustrated the complexity of countryside, full of tension and imagination, the metaphor of the high-tension line is also very important in the rural study.

This paper employs Shanghai's Heping Village as an example to analyze the status and driving forces of rural construction, and includes five sections. Following the introduction, the details of the research area and data and research methods are shown. The third section analyzes the status of rural construction in the case area. The fourth section discusses suggestions for future rural construction in Heping Village, and the final section summarizes the conclusions and contributions of this article.

10.2 Study Area and Research Methodology

10.2.1 Study Area

Heping Village of Shanghai is located on the west side of Wujing Town (Fig. 10.1). It is bordered on its east by the Yingtao River, its south by Jianchuan Road, its west by the Danshui River, and its north by the Yutang River. South Lianhua Road bisects it horizontally. In 2016, the village had 2554 households, with a resident population of 7710, of which 5707 were migrants (Fig. 10.2). There were 215 employees in its primary industry, 3680 workers in its secondary, and 1259 in the tertiary. In recent years, the local population number has barely changed; however, the migrant population has gradually been decreasing. This decrease in the migrant population is due in large part to the increase of housing rents caused by the aesthetic improvements of Heping Village, forcing some people to move to low-cost housing.

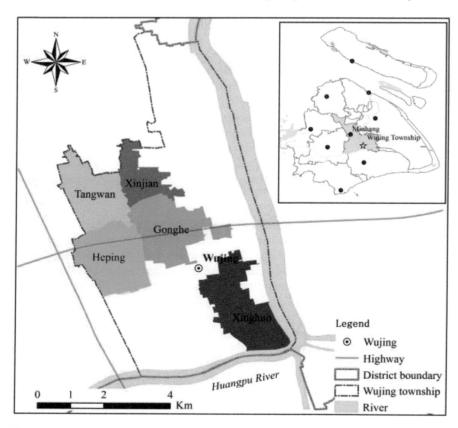

Fig. 10.1 Location of Heping village in Shanghai

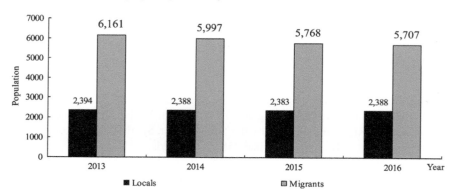

Fig. 10.2 Heping Village's population changes from 2013 to 2016. *Source* SMDBS [32–35]

10.2 Study Area and Research Methodology

Shanghai has promoted this "beautiful rural construction" since 2016. Heping Village became the pilot village representing the "beautiful countryside construction of Wujing". Its construction type was designated as an "ecological village," which requires adherence to the highest construction standards in contemporary rural construction, an expensive proposition. Over the past two years, changes in investments for similar social undertakings in Heping Village have greatly increased, from 9.9215 million yuan in 2015, to 27.28 million yuan in 2016. In 2015, the funds were mainly used to improve non-compliant buildings that were part of environmental projects; in 2016, the funding increases were invested in transforming the village on the above basis.

As a globalized megalopolis, Shanghai's development is largely influenced by national policies. In 1984, the Central Committee of the Communist Party of China (CCCPC) decided to expand 14 coastal cities, including Shanghai. In 1990, the CCCPC and the State Council announced a plan to expand and develop the Pudong district, using many specific policies that included strengthening infrastructure construction and investing in a number of large projects. Since then, Shanghai's economy and urban construction have developed at an unprecedented rate. Shanghai's center has gradually expanded from downtown to the Pudong New Area. Meanwhile, Wujing Town, an old industrial base and suburban town of Shanghai, has been neglected for years. Wujing's economic and social development thus lags far behind the average development level of Shanghai. Heping, as one of Wujing's villages, is also affected by its economic and geographical conditions. There are wide developmental gaps between the globalized city center, the New Area, and the marginalized countryside, which affect the shape of the countryside under multiple high-tension lines.

Heping Village serves as a typical case study of rural construction in China for the following reasons. First, from a global perspective, Shanghai is one of the largest metropolitan areas in the world and is representative of China's rapid urbanization, and Heping's rural construction is greatly affected by both China's and Shanghai's strategies and responses to this rapid urbanization. Second, Shanghai is an important economic, transportation, technological, industrial, and financial center in China, and capital plays an important role in the process of rural construction. Third, Heping is located in Wujing Town, Minhang District, and, as one of the suburbs closest to downtown; its rural construction involves many factors. Finally, Heping was built as a model village by the town government, receiving the largest portion of funding and support compared to other villages of Wujing Town, yet there are many houses under multiple high-tension lines. Heping Village is geographically small but reflects the interactions between different spatial scales and governmental power and capital in the process of rural construction.

10.2.2 Research Methodology

Our integrated methodology includes three parts. First, the Minhang Statistical Yearbook [32–35] and the Statistics Bulletin on Urban–Rural Construction [25] are the official sources for our data statistics. The Statistical Yearbook represents a collection and arrangement of basic data from Wujing's villages from 2014 to 2017, used in order to ascertain generalized information about the villages. The Statistics Bulletin on Urban–Rural Construction provides a background on rural construction and illustrates the allocation of investment in rural construction, aiding in understanding the overall situation (see Table 10.1).

Second, the field investigation includes questionnaires (105 valid questionnaires in 120 issued questionnaires) interviews (Table 10.2). These questionnaires mainly examined the residents' social participation and sense of belonging. The interviews were a combination of pre-designed questions and semi-structured interviews, conducted through 20–30 min conversations with different individuals, such as local people, migrants, and village officials, representing different social classes' different views on rural construction. Through this qualitative research, the researchers aimed to understand individual situations and community from the subjects' point of view, by learning about people's social and material circumstances, and their experiences, perspectives, and histories [10, 31].

Finally, the authors and researchers of this paper participated in a rural construction project called the "beautiful countryside construction of Wujing" project. This project began in December 2016, and ended in May 2017, during which time the researchers used their theoretical knowledge to plan and establish relevant countermeasures. It is

Table 10.1 An integrated methodology on rural studies

Methods	Sources	Objectives	Forms of representation	Roles of authors
Official Data	Statistical yearbook	Generalized	Graphs	User
Fieldwork	Questionnaire and interview	Individual	Interpretation	Researcher
Planning	Rural planning	Balance individual and generalized data	Photos and text	Planner

Table 10.2 Participant information from the valid samples

Item	Category and proportion			
Gender	Male: 53.3%		Female: 46.7%	
Age distribution	9.5% (25)	26.7% (26–40)	35.2% (41–55)	28.6% (56)
Type of registered permanent residence	Agricultural registered permanent residence: 70.5%		Non-agricultural registered permanent residence: 29.5%	
	Local	Migrant	Local	Migrant
	40%	30.5%	17.2%	12.3%

noted that we used the theory of critical methodology—the production of space—in this planning. Our participation in the planning helped balance the relations between the generalized and the individual, between the characteristic and the common, and between participants and researchers. The integration of these three methods can comprehensively reflect the reality of rural redevelopment and construction.

10.3 The Countryside Under Multiple High-Tension Lines

10.3.1 Physical High-Tension Lines

In recent years, Heping Village has experienced several instances of new rural construction. The village has implemented new plans: river regulations, road rebuilding, demolishing illegal structures; a factory closing; and even altering certain residents' living spaces. However, in the investigation, some resident's houses are still located in the path of the high-tension lines, as well as very close to the high-pressure tower (Fig. 10.3), although it is extremely dangerous and illegal. Generally, an inhabitant's choice of residence is affected by individual, household, and/or contextual factors [50]. When asked about his reasons for living in Shanghai, a respondent said.

> We moved here in 2013, originally living on South Lianhua Road and then the government gave us the house here. Now, half of the houses here are rented out to migrant residents.

Another worker said,

> I'm a migrant and renting here. Although the environment here is not ideal, I feel pretty good because of my nearby work.

Due to changes to the eastern part of South Lianhua Road to accommodate "Wujing Country Park" as per the new plan, the government moved the inhabitants of the east side to the west, though the living area was located under the high-tension lines. The purpose of the government's remodel of the park was purportedly to

Fig. 10.3 Houses under high-tension lines. *Source* Authors' photos

improve the residents' physical environment; ironically, the problem of dangerously located housing was not on the government's agenda. The residents here worried that their houses would be struck by lightning due to their proximity to the high-tension lines; however, the government did not discuss resolving this situation.

10.3.2 Power High-Tension Lines

Power in China is manifested primarily in politics, policies, and systems. For example, rural construction always begins with policies set forth by the central government. In the planning of the "beautiful countryside construction of Wujing," Heping Village was considered to be the key developmental object. The town's party secretary emphasized Heping Village's status as the best and "very typical" part of the plan. In her discussion of the "beautiful countryside construction of Wujing," she said:

> From the current situation, Heping Village can be expected to exist for a long time; it represents the "new rural construction" as does Xinjian Village. Heping Village is in a relatively good location. We have completed the weight of the work; people should pay more attention to the construction of Heping Village. This village is very typical: after the demolition, the village economy has transformed; the foundation has been rebuilt; there has been a typical transformation of the river; the village is well known as an agriculture center of Shanghai; it is a good human environment. We intend to transform the forest belt that is under the high-tension tower into the Wujing Country Park, and then the countryside will become a real ecological village. Thus, the rural construction of Heping Village, carried out through intense planning, will provide guidance to ensure that other villages are in accordance with the basic principles of reconstruction.

In her view, Heping Village has many advantages compared to other villages.

Heping Village's construction is indeed neater than other villages' (Fig. 10.4). Referring to this, a local man said:

> After the river dredge, the environment became better. The river was dark before and we did not want to go out for a walk; the river … smelled bad in the summer. Now the river is clean and there is a corridor along the river, where I walk after dinner. I feel that my life is full of comfort and better than my original life in the city.

Heping Village's river is now very clean, due to the new rural construction, and residents reported enjoying walking and participating in other recreational activities in its vicinity. It has become a "typical" river in the town, an accomplishment commended by the government. However, due to this improvement, rent prices have increased sharply, prompting the following complaint from a migrant:

> I do not know whether it is possible to control the rental housing costs. Last year's price increased crazily, almost doubled; coupled with the cost of living, we cannot make ends meet and we have no savings … What I wish most is the department could care about us (migrants), because the local people of Shanghai do not care about migrants at all.

10.3 The Countryside Under Multiple High-Tension Lines

Fig. 10.4 Comparison of river conditions between Heping Village (left) and Tangwan Village (right). *Source* Authors' photos

Heping Village does have an aesthetic advantage, but its actual living conditions are less than ideal. Along with concerns over escalating rent, residents are also concerned about their safety. Additionally, Heping Village lacks community facilities for its residents. In actuality, Heping Village offers no greater practical advantages than any other village.

In the "beautiful countryside construction of Wujing" project, village construction is divided into three types: "ecological," "livable," and "clean and tidy" (Table 10.3). These three terms represent different construction standards: an ecological village will undergo complete environmental remediation and ecological restoration, representing the optimal living environment; the livable village is focused on beautifying and cleaning, mainly to enhance the village environment; and clean and tidy, the lowest level of rural construction, is the standard used for those villages that may be demolished in the next few years. Since governmental officials perceived Heping Village as having multiple advantages, they chose it as the pilot site for ecological village construction. In the whole process of rural construction, whether at Heping Village or other villages, the development plan is determined entirely by government departments, exclusive of the villagers' views.

Some participants stated that the village committee, an organization of village affairs, did not fight for the interests of the villagers. A local elder complained that:

> Due to the demolition of buildings, the original storage of gas cylinders was torn down; gas cylinders can now only be placed in the living room, and we are nervous every day. The village committee has no practical effect; some in the group just pretend to ask what we need. In some case, such as the demolition of the housing with the land certificate, they did not give us a solution. The government did not come forward to solve nor ask us to sign the consent of the guarantee; can we not sign it? After all, the people still have to obey the government, or else we will suffer.

Table 10.3 Criteria for the "beautiful countryside construction of Wujing" project

Serial number	Construction content	Ecological type		Livable type	Clean and tidy type
1	Road construction	✓		✓	✓
2	Bridge construction	✓		✓	
3	Parking lot construction	✓		✓	
4	Treatment of domestic sewage facilities	✓		✓	✓
5	Regulation of river	✓		✓	✓
6	Construction of water supply in village house	✓			
7	Environment improvement in the front and back of the house	✓		✓	
8	Renovation of village public environment	✓		✓	✓
9	Forestation of village	✓		✓	✓
10	Arrangement of dustbin and toilet	✓		✓	
11	Construction of agricultural production facilities	✓		✓	
12	Construction of ecological agriculture	✓			
13	Villager autonomy	✓		✓	✓
14	Construction of public service facilities	Rural community affairs agency room		✓	
15		Senior center		✓	
16		Clinic		✓	
17		Cultural activity room		✓	
18		Fitness trails, sports facilities, fitness spots		✓	
19		Extending radio and TV broadcasting coverage to every village project		✓	

(continued)

10.3 The Countryside Under Multiple High-Tension Lines

Table 10.3 (continued)

Serial number	Construction content	Ecological type	Livable type	Clean and tidy type
20	Curb illegal land use and illegal construction		✓	✓

Data source authors' project

10.3.3 Capital High-Tension Lines

According to the Statistics Bulletin on Urban–Rural Construction, the total investment in village and town construction in 2016 was over 1.59 trillion yuan. According to the geographical division, the amount of yuan invested, in billions, was as follows for the specified areas: built-up areas of organic towns, 682.5; built-up areas of townships, 52.4; special areas of the town and township, 23.8; villages, 832.1. These amounts respectively account for total investments of 42.9%, 3.3%, 1.5%, and 52.3%. According to reported usage (Fig. 10.5), the investment in building construction was 1188.2 billion yuan, and the investment in municipal utilities construction was 402.6 billion-yuan, accounting for 74.7% and 25.3% of total investment respectively. The investment in municipal utilities construction was mainly used for road and bridge construction, drainage, landscaping, and environmental maintenance. The construction funds allotted for villages and towns throughout the country are mostly used for residential construction; however, this was not in support of actual residential living. Not only are the inhabitants' residences unsafe, but the allocation of funds is not balanced.

At Heping Village, the construction of public service facilities accounted for the largest proportion of allocated funds, which included the cost of demolishing illegal buildings (Fig. 10.6); however, the cost of dismantling these buildings accounted for

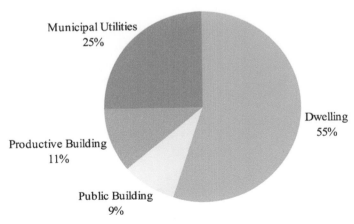

Fig. 10.5 2016 investment structure for village and town construction. *Source* MOHURD [25]

only about 9.5% of these allocated funds. It is noteworthy to observe that "illegal buildings" are defined as buildings built by villagers; the definition does not address whether a structure was built illegally under a high-pressure tower. This substantiates the villagers' assertion that their housing issues have received little attention. Greater focus has been given to the creation of comprehensive environmental regulations, reclamation of land and forestation, as well as municipal infrastructure, which include the construction of village roads. The sum of the construction funds of the comprehensive environmental regulations, reclamation of land and forestation, and municipal infrastructure has already exceeded the construction funds allocated for public service facilities. In fact, housing reform, community centers, and fitness facilities are the residents' main concerns, yet little funding has been earmarked for these requests. The larger amount of money has been invested in roads, forestation, and environmental improvement, in order to achieve the more superficial aim of creating an aesthetically pleasing village.

In the planning process, the government stressed three points—"infrastructure first," "including all villages," and "sustainable development"—which seemed to be aimed at the overall improvement of Wujing. However, the project's actual implementation is inconsistent with this declaration of its spirit. For example, due to Heping Village's stronger pre-existing foundations and superior circumstances as compared to other villages, most of the project's funds (about 30 million yuan) were invested in Heping Village's construction, despite its smaller population, in order to highlight the construction project and satisfy Wujing's planning (Fig. 10.7). While Xinghuo Village's population is similar to that of Heping Village, it received only about 2 million yuan to invest in its construction, in large part because it may be demolished

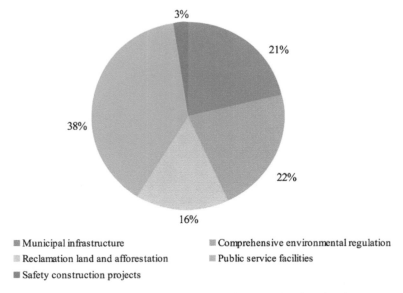

Fig. 10.6 The investment structure of Heping Village. *Source* Authors' project data

10.3 The Countryside Under Multiple High-Tension Lines

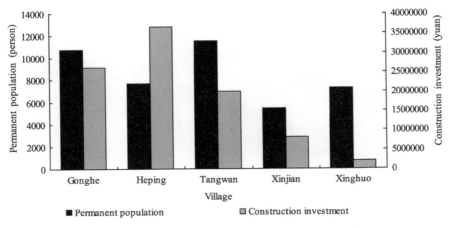

Fig. 10.7 Construction investment and population comparison of Wujing's villages. *Source* SMDBS [35] and authors' project data

next. Gonghe Village and Tangwan Village have a greater number of permanent residents than Heping Village, yet also received less to invest. The people-oriented urbanization is often talked by officials; however, in actuality, "people" has often been ignored, and the majority of villagers receive less attention and investment. Villagers' actual lives are under the control of capital and political interests: when the countryside in a given rural area is perceived as desirable, it becomes a "model" for the government and receives increased investment and development opportunities. Although Heping Village has received a greater portion of the apportioned construction funds, it is clear that from the criteria for "the beautiful countryside construction of Wujing" project, the focus of the construction is more on cleaning and beautifying the village, rather than promoting community culture. Further establishing evidence of this contradiction, the ecological-type villages are the only type specified to include cultural construction.

10.3.4 Psychological High-Tension Lines

China primarily addresses financing of rural area projects through top-down policies, rather than from the residents' perspectives. The contents of the new rural construction policies issued in the process of urbanization are mostly basic construction standards. Usually, in order to most rapidly achieve new rural construction goals, government construction projects are directly aligned with the stated standards, and do not consider the wishes, building ideas, and needs of a village's residents. Therefore, residents are forced to "illegally" construct certain types of structures and spaces in order to meet their otherwise overlooked objectives. Despite a lack of provision for such structures in the new construction project standards, many village residents

Fig. 10.8 The empty Senior Center (left) and Women's Center (right). *Source* Authors' photos

in Heping Village have established storage rooms and vegetable gardens that occupy public spaces. The residents have a decreased sense of belonging, and they do not feel that the village is their "home." One local resident said:

> The east of the village is to be modified and it is said that the garden will be planned for viewing. However, we are retired; the pension is only 1000 yuan, which is not enough for the usual living expenses. If there is land then we can sell vegetables for the family. The government is always doing these impractical things; the garden is a surface project that has no use for us.

There is scant space for public interaction in the village, though there are some public institutions, such as the Senior Center, and the Women's Center (Fig. 10.8), but these have received less government funding. At the same time, residents' utilization of these institutions is extremely rare because the structures as they are do not meet the residents' needs. A young migrant said:

> There is rarely any activity here. The old activity room (Senior Center) and the Women's Center do not have much to do with us and even the locals do not go there. It is already full of dust; there is no one there.

Meanwhile, the development of network technology has changed the common values of rural life, reducing residents' emotional connections to the village, and even causing social separation [13]. When we asked a junior high school student if he participated in any recreational activities, he replied,

> I usually stay at home and play on my cell phone; I neither have other recreational activities nor have heard of village activities.

The number of resident activity centers is limited, useful for and enjoyed by only a small number of people. The obsolescence of the facilities and the absence of organized community activities contribute to the tedium complained of by residents. Infrequent interpersonal communications between villagers also exacerbates such social problems.

The insufficiency of investment in the construction of community facilities that could serve to improve the quality of the residents' lives is troubling. The village residents feel disconnected from rural construction, and powerless regarding its development. They do not feel that is for their benefit, and thus they decline to participate in the process.

Additional estrangement also exists between locals and migrants. Little communication occurs between them, and the manner in which they receive welfare differs. The secretary of Heping Village said:

> For the autonomous management of the village, we give priority to Heping villagers because many villagers here are now unemployed and the elderly are not able to find employment. The cleanliness and security of our villages are also given priority for the villagers because they need to be healthy and safe.

The village officials are more inclined to protect the interests of local people at the expense of migrants, a situation that obviously frustrates migrants.

10.4 Discussion

From the case of Heping Village, it can be seen that different stakeholders have different reasons for and expected outcomes of village construction. Villagers, both local and migrant, wish to feel an emotional connection to their village, brought about perhaps by enhanced interpersonal connections. Such connections may be fostered by the construction or renovation of community centers, staffed by trained personnel who are adept at promoting community activities. Residents are also interested in increasing their income and/or having their rent lowered to relieve the pressure of daily life. Additionally, despite the village's proximity to schools, migrants still lack educational opportunities. These elements should be considered foremost by the government during a village's construction. The government, alternatively, seemingly views village construction projects as opportunities to display its work, rather than bettering its residents' lives by listening and responding to their needs.

Even village committees, which act as liaisons among villagers and between villagers and the government, seemingly fail to assist villagers. The distinctly different ways in which committees treat locals and migrants is not fair. The committees seem more closely aligned with the government's purposes, rather than concerning themselves with expressing the villagers' needs.

In the case of Heping Village's multiple high-tension lines, we can learn that the pleasing aesthetic nature of this rural construction did not improve the residents' quality of life. The local stakeholders' (i.e., village residents) participation in rural construction plays an important role in promoting sustainable rural development [11, 26, 44]. Hence, it is necessary for the villagers, village officials, and government departments to work together to change the status of Heping Village.

There is a variety of ways in which villagers could insert themselves into rural construction that would address several of their concerns. For example, they might consider forming organizations to carry out community construction. In terms of production, they could unite local enterprises or other likeminded people to establish industries related to local agriculture, culture, or landscape, in order to improve their income and attract the government's attention. In terms of their daily lives, villagers should act, instead of merely protesting or complaining. A village group could be formed, the purpose of which might be to obtain other villagers' opinions regarding construction, which could then be provided as feedback to the village committee or higher government departments. These measures would enhance public participation in community construction and strengthen the villagers' sense of responsibility for and belonging to the community.

For these reasons, local governments should not ignore villagers' needs and opinions. There are three steps the government can take in order to ensure that villagers are included in rural construction. First, in the pre-planning stage, the government must seek the villagers' input. This information could be obtained through interviews, questionnaires, or at public meetings. Second, during the planning stage, the government must prioritize projects related to the villagers' quality of life and personal safety. For example, the government must address the fact that some residents of Heping Village live under high-pressure towers: this is known to be extremely dangerous, as discussed above. However, the government seems either unaware of, or unconcerned with, this, and instead continues directing funds toward beautification. Finally, once the planning is complete, and due to the ongoing nature of village construction, the government must closely attend to any potential problems of the construction, rather than merely watching the process.

The production of space in the countryside is a process that is monitored by different departments. For example, the construction of Heping Village is under the management of several entities: the Shanghai Municipal Government, the Minhang District Government, the Wujing Town Government, and the Heping Village Committee. The Shanghai Municipal Government, from the overall perspective of the city, pays more attention to the development of downtown and the Pudong New Area; Minhang District is not its concern. The development of different towns in Minhang District is also accomplished differently. For example, Wujing Town is an advantageously located traditional industrial city, but is in decline; its village committee fights primarily for its interests. The result of urbanization is production of space, caused by the combination of power and capital. Ultimately, the result of urbanization is that residents are physically and psychologically marginalized, as if they were living under multiple high-tension lines.

Compared with the previous "new socialist countryside construction" [18], the "beautiful countryside construction" standards seem to be more concerned with the transformation of rural landscape. On the one hand, budgets allocated to rural construction can be used to fund construction of already-built environments, such as residents' dwellings. On the other hand, the distribution of funds in different villages is uneven and unfair, and it is not applied toward the most urgent rebuilding needs of the village's residents. As a result, some villages remain under multiple high-tension lines.

10.5 Conclusion

Although there has been progress in China's rural development, the villages in China are confronted with a crisis. Heping Village is a typical case that could reflect the aspects of many rural areas, especially those in megacities. Even though it is located in Shanghai, one of the biggest and most modern cities in China, Heping Village is under multiple high-tension lines. Physically, the houses and everyday life of the villagers are very close to the high-tension lines and towers, which are extremely dangerous but have not been taken seriously by the government. Furthermore, the high-tension lines from capital and power, which are produced by the pattern of production of space in the process of China's urbanization, have eroded and destroyed the living environments of the residents. Rural construction funds are unfairly distributed by the government, and only focus on infrastructures and so-called "good-looking" villages, overlooking the community and real needs from the residents. As a result, psychological high-tension lines are produced because the villagers have no feeling of belonging to their community and villages.

The top-down mode of rural planning and construction also leads to the dilemma of rural development. The movement of beautifying the countryside has enhanced rural development to some extent; however, it lacks bottom-up initiatives and driving forces [12]. The construction of rural community requires integration of multiple agents including the government, residents, planners, developers, and other organizations. The roles of the different main actors need to be reconsidered. In future planning and policy making, villagers should actively show a willingness to participate in rural construction, the village committee should undertake the responsibility of facilitating communication between villagers and the government; and the government should prioritize voices from the villagers and involve them in the full process of rural construction. Meanwhile, more village representatives should be elected and encouraged to vote for rural plans and to increase the participation of the residents.

The methodology and theory of rural research needs to be renewed as the rural world itself changes. On the one hand, research that uses multi-method integration is important and necessary now that real rural issues are complicated and multi-faceted; cross-disciplinary efforts and collaborations are critical. On the other hand, some new methods, such as artistic expression, can be explored and used to more fully explain and reconstruct the developmental process of countryside. Metaphor is a key and interesting method that can help to discover and illustrate real rural problems. In our paper, the metaphor of high-tension lines is critical and expressive. It is worth exploring how other artistic means can be applied to the explanation and even the reconstruction of rural society and landscape.

References

1. Buser M (2012) The production of space in metropolitan regions: a Lefebvrian analysis of governance and spatial change. Plan Theory 11(3):279–298

2. Chen J (2007) Rapid urbanization in China: a real challenge to soil protection and food security. CATENA 69(1):1–15
3. Chan KW (2010) Fundamentals of China's urbanization and policy. China Rev 10(1):63–93
4. Chen M, Liu W, Lu D (2016) Challenges and the way forward in China's new type urbanization. Land Use Policy 55:334–339
5. Chen X, de'Medici T (2010) Research note—the "instant city" coming of age: production of spaces in China's Shenzhen special economic zone. Urban Geogr 31(8):1141–1147
6. Cloke P (2003) Country visions. Prentice Hall, London
7. Friedmann J (2006) Four theses in the study of China's urbanization. Int J Urban Reg Res 30(2):440–451
8. Frisvoll S (2012) Power in the production of spaces transformed by rural tourism. J Rural Stud 28(4):447–457
9. Halfacree K (2007) Trial by space for a "radical rural": introducing alternative localities, representations and lives. J Rural Stud 23(2):125–141
10. Kvale S (1994) Interviews: an introduction to qualitative research interviewing. Theory Psychol 19(2):267–270
11. Lang W, Chen T, Li X (2016) A new style of urbanization in China: transformation of urban rural communities. Habitat Int 55:1–9
12. Li Y, Westlund H, Zheng X, Liu Y (2016) Bottom-up initiatives and revival in the face of rural decline: case studies from China and Sweden. J Rural Stud 47:506–513
13. Lin G, Xie X, Lv Z (2016) Taobao practices, everyday life and emerging hybrid rurality in contemporary China. J Rural Stud 47:514–523
14. Liu Y, Lu S, Chen Y (2013) Spatio-temporal change of urban–rural equalized development patterns in China and its driving factors. J Rural Stud 32:320–330
15. Liu J, Liu Y, Yan M (2016) Spatial and temporal change in urban-rural land use transformation at village scale—A case study of Xuanhua district, North China. J Rural Stud 47:425–434
16. Liu Y, Yu L, Chen Y, Long H (2010) The process and driving forces of rural hollowing in China under rapid urbanization. J Geog Sci 20(6):876–888
17. Long H, Woods M (2011) Rural restructuring under globalization in eastern coastal China: what can we learn from Wales? Manitoba Ministry Agric Food Rural Initiatives 6(1):70–94
18. Long H, Zou J, Pykett J, Li Y (2011) Analysis of rural transformation development in China since the turn of the new millennium. Appl Geogr 31(3):1094–1105
19. Long H, Li Y, Liu Y, Woods M, Zou J (2012) Accelerated restructuring in rural China fueled by "increasing vs. decreasing balance" land-use policy for dealing with hollowed villages. Land Use Policy 29(1):11–22
20. Long H, Liu Y, Hou X, Li T, Li Y (2014) Effects of land use transitions due to rapid urbanization on ecosystem services: implications for urban planning in the new developing area of China. Habitat Int 44:536–544
21. Long H, Tu S, Ge D, Li T, Liu Y (2016) The allocation and management of critical resources in rural China under restructuring: problems and prospects. J Rural Stud 47:392–412
22. Marsden T (1996) Rural geography trend report: the social and political bases of rural restructuring. Prog Hum Geogr 20(2):246–258
23. Marsden T, Lowe P, Whatmore S (1990) Rural restructuring: Global processes and their responses. rural restructuring global processes & their responses. David Fulton, London
24. McGee TG (2009) Interrogating the production of urban space in China and Vietnam under market socialism. Asia Pac Viewp 50(2):228–246
25. Ministry of Housing and Urban-Rural Development of the People's Republic of China (MOHURD) (2016) Statistics Bulletin on urban and rural construction. Retrieved from http://www.mohurd.gov.cn/xytj/tjzljsxytjgb/tjxxtjgb/201708/t20170818_232983.html
26. Molden O, Abrams J, Davis EJ, Moseley C (2017) Beyond localism: the micropolitics of local legitimacy in a community-based organization. J Rural Stud 50:60–69
27. Nasongkhla S, Sintusingha S (2013) Social production of space in Johor Bahru. Urban Stud 50(9):1836–1853

References

28. Nelson PB (2001) Rural restructuring in the American west: land use, family and class discourses. J Rural Stud 17(4):395–407
29. Qian W, Wang D, Zheng L (2016) The impact of migration on agricultural restructuring: evidence from Jiangxi province in China. J Rural Stud 47:542–551
30. Qin H, Liao T (2016) Labor out-migration and agricultural change in rural China: a systematic review and meta-analysis. J Rural Stud 47:533–541
31. Ritchie J, Lewis J (2003) Qualitative research practice: a guide for social science students and researchers. Sage, London
32. Shanghai Minhang District Bureau of Statistics (SMDBS) (2014) The Minhang statistical yearbook. China Statistics Press, Beijing. [上海市闵行区统计局. 2014. 闵行统计年鉴. 北京: 中国统计出版社]
33. Shanghai Minhang District Bureau of Statistics (SMDBS) (2015) The Minhang statistical yearbook. China Statistics Press, Beijing. [上海市闵行区统计局. 2015. 闵行统计年鉴. 北京: 中国统计出版社]
34. Shanghai Minhang District Bureau of Statistics (SMDBS) (2016) The Minhang statistical yearbook. China Statistics Press, Beijing. [上海市闵行区统计局. 2016. 闵行统计年鉴. 北京: 中国统计出版社.]
35. Shanghai Minhang District Bureau of Statistics (SMDBS) (2017) The Minhang statistical yearbook. China Statistics Press, Beijing. [上海市闵行区统计局. 2017. 闵行统计年鉴. 北京: 中国统计出版社]
36. Tao R, Xu Z (2005) Urbanization, rural land system and migrant's social security. Econ Res J (12):45–56. [陶然, 徐志刚. 2005. 城市化、农地制度与迁移人口社会保障——一个转轨中发展的大国视角与政策选择. 经济研究, (12): 45–56]
37. Tian Q, Guo L, Zheng L (2016) Urbanization and rural livelihoods: a case study from Jiangxi province, China. J Rural Stud 47:577–587
38. Tu S, Long H (2017) Rural restructuring in China: theory, approaches and research prospect. J Geog Sci 27(10):1169–1184
39. Whatmore S, Munton R, Marsden T (1990) The rural restructuring process: emerging divisions of agricultural property rights. Reg Stud 24(3):235–245
40. Whatmore S (1993) Sustainable rural geographies? Prog Hum Geogr 17(4):538–547
41. Wilson J (2013) The Urbanization of the countryside depoliticization and the production of space in Chiapas. Lat Am Perspect 40(2):218–236
42. Wilson OJ (1995) Rural restructuring and agriculture-rural economy linkages: a New Zealand study. J Rural Stud 11(4):417–431
43. Woods M (2005) Rural geography: processes, responses and experiences in rural restructuring. Sage, London
44. Woods M (2008) Social movements and rural politics. J Rural Stud 24(2):129–137
45. Woods M (2009) Rural geography: blurring boundaries and making connections. Prog Hum Geogr 33(6):849–858
46. Woods M (2009) Rural geography. In Kitchin R, Thrift N (Eds) International encyclopedia of human geography. Elsevier, Oxford
47. Woods M (2010) Performing rurality and practising rural geography. Prog Hum Geogr 34(6):835–846
48. Woods M (2012) Rural geography III: rural futures and the future of rural geography. Prog Hum Geogr 36(1):125–134
49. Yang X (2013) China's rapid urbanization. Science 342(6156):310
50. Yang C, Xu W, Liu Y, Ning Y, Klein KK (2016) Staying in the countryside or moving to the city: the determinants of villagers' urban settlement intentions in China. China Rev 16(3):41–68
51. Yao Y, Si X, Ye W (2017) The spatial transformation mechanism of Beijing Songzhuang Cultural and Creative Industry Zone: a perspective of production of space. In International conference on logistics, informatics and service sciences, pp 1–4
52. Ye C, Chen M, Chen R, Guo Z (2014) Multi-scalar separations: land use and production of space in Xianlin, a university town in Nanjing, China. Habitat Int 42:264–272

53. Ye C, Chen M, Duan J, Yang D (2017) Uneven development, urbanization and production of space in the middle-scale region based on the case of Jiangsu province, China. Habitat Int 66:106–116
54. Yep R, Forrest R (2016) Elevating the peasants into high-rise apartments: the land bill system in Chongqing as a solution for land conflicts in China? J Rural Stud 47:474–484
55. Yu A, Wu Y, Zheng X, Zhang, Shen L (2014) Identifying risk factors of urban-rural conflict in urbanization: a case of China. Habitat Int 44:177–185
56. Zhang Y, Li X, Song W, Zhai L (2016) Land abandonment under rural restructuring in China explained from a cost-benefit perspective. J Rural Stud 47:524–532
57. Zhao S, Da L, Tang Z, Fang HS, Fang J (2006) Ecological consequences of rapid urban expansion: Shanghai, China. Front Ecol Environ 4(7):341–346
58. Zhu Z, Zheng B (2012) Study on spatial structure of Yangtze River delta urban agglomeration and its effects on urban and rural regions. J Urban Plann Dev 138(1):78–89

Chapter 11
The Essence of Production of Space and Future of Urbanization

Abstract The theory of production of space has been the focus of multidisciplinary research since the 1970s to the present. From Lefebvre to Harvey, Soja, Castells, Smith, and the contemporary critical urban theorist Neil Brenner, the European and American enthusiasm for theoretical and empirical research on production of space has continued unabated. This theory has been introduced in Chinese academic circles since the 1990s, and it has received particular attention and formed a research upsurge in the past 10 years.

11.1 Three Characteristics of the Theory of Production of Space

The theory of production of space has three main characteristics: social-spatial dialectics, multi-scale synthesis and critical construction. This provides new ideas and perspectives for the research and practice of urbanization, so it has been paid more attention by scholars in different fields. The premise of understanding and grasping the theory of production of space is to grasp the social-spatial dialectics, which emphasizes that the social process and the spatial form are not two different things (or processes), nor is it a relationship that one includes or reflects the other, but a relationship that is blended and indistinguishable. The interaction between society, space and time (in a broad sense) eventually forms a ternary dialectical relationship.

In addition to socio-spatial dialectics, production of space has two other characteristics: multi-scale synthesis and critical construction. The so-called multi-scale synthesis means that it can scale up or down from small (such as "body" in postmodern geography) to large (global scale) according to the scale of geographical units and can be divided into environment, territory (land), (narrowly) society, culture, sexual (gender), and other contents according to the specific connotation, and these scales are intertwined. Critical construction means proposing new theoretical frameworks for planning and practice while revealing the complex nature of unjust urbanization. Social-spatial dialectics, multi-scale synthesis and critical construction together constitute the three dimensions of production of space theory, providing theoretical guidance for China's urbanization research.

11.2 Production of Space Is the Driving Force of Urbanization

The theory of production of space is closely related to the reality of rapid urbanization in China. Since the 1990s, China's urbanization and urban space have undergone drastic changes, and many spatial problems caused by the rapid urbanization model that emphasized land expansion and economic growth in the past have begun to emerge. Spatial injustices at different levels or scales have affected the healthy development of urbanization. After entering the twenty-first century, especially in the context of economic and social transformation and the new era, it has become an urgent task to reflect on China's urbanization development model in combination with the national strategy of new-type urbanization.

China's urbanization rate increased rapidly from 17.92% in 1978 to 64.72% in 2021, which took hundreds of years in developed countries in only a few decades and overlaps with globalization, industrialization and informatization. Unprecedented space–time compression has created an unprecedented urbanization landscape. Looking back at history, in the early years of reform and opening up, China's urbanization took economic development as its foremost goal. The emphasis on the efficiency of social and economic development makes urbanization characterized by super-large-scale land use and spatial expansion. The production of various special economic zones, development zones, university towns, new urban districts, and other spaces is manifested in the rapid evolution of social economy and urbanization characterized by the construction and renewal of urban spaces. As China's urbanization is more and more integrated into the process of globalization, the unprecedented flows of people, logistics, capital, information and their ultra-high-speed flows make urbanization present a complexity of intertwined elements: the information-based and networked society reduces the barriers of geographic distance and strengthens the connection between local and global; big data, intelligence and advanced e-commerce make life more convenient and exacerbate the complex evolution of social space; globalization drives the transfer and agglomeration of capital around the world and reshapes the regional development patterns. Whether it is the global urban network space at the macro scale or the everyday life space of urban and rural residents at the micro scale, vast and unpredictable changes have taken place. Similarly, these different scales of space also carry and dominate the time and space from global urbanization to daily life. The uniqueness of China's urbanization is actually subordinate to this general trend of global change. "China" has become a typical "sample" of the evolution of urbanization in the contemporary world, and to some extent, "China" is increasingly represented as the "world". Therefore, although the empirical and case studies in this book are mainly aimed at the different scales of urbanization in China, in theory, it attempts to go beyond the "exceptionalism" of China's urbanization. The production of space just so happens to provide an excellent theoretical weapon for this study.

In a nutshell, China's urbanization is governed by the logic of production of space, which is particularly evident under the trend of globalization. The interaction and influence of urbanization and production of space is a complex process interwoven at multiple scales. The natural spaces under the control of production of space have begun to be transformed into social spaces, and now there is no natural space in the pure sense. In terms of the driving forces of social spaces, capital agglomeration leads to imbalanced geographical space and sharply differentiated social spaces, and power gives birth to special spatial organization forms such as urban agglomerations, metropolitan circles, high-tech zones, new areas, and gated communities. The combined action of the two forces dominates the individual spaces of everyday life.

From macro to micro, the interactions of multiple scales and elements are difficult to separate. Especially in the past 20 years, the new driving forces for the evolution of China's urbanization are power and capital, and their interaction with class has reshaped the social spaces of China's urbanization. Capital represents aspects of the economy, such as foreign direct investment, fixed asset investment (including real estate investment). Power is mainly embodied in top-down strategy, policy, planning and all kind of discourses. Class refers to the different groups and individuals in a society. Power shapes space with the help of national strategies, planning and policies, and capital uses financial tools to redistribute resources such as land and produce spaces. Rapid urbanization driven by land finance has resulted in uneven geographic development. The production of space is accompanied by the reorganization of social relations, and between this process and the urbanization process of transformation from rural to urban space is a relationship of mutual shaping and reshaping. The three elements of capital, power and class operate together in urbanization, and the spatial transformation of urbanization produces and affects these three elements. In this way, all kinds of abstract and physical spaces can be continuously shaped and reshaped. This is the core connotation of production of space that we redefined at the beginning: the production of space means the space is constantly reshaped by the political, economic, and social factors that are mainly embodied by power, capital, and class.

11.3 Multi-scalar Intertwined Production of Space

The most important and also the most difficult task for understanding and using the theory of production of space research is methodology. Methodologically, we should be problem-oriented to abandon disciplinary boundaries for interdisciplinary and multi-methodological integration. Specifically, we should give full play to the characteristics of different disciplines. The true problem-oriented is based on the research objects or groups and different scales, starting from specific socio-spatial issues to realize the integration of interdisciplinary approaches. For example, big data and quantitative analysis methods can be used on a large scale, while methods such as questionnaire interviews and fieldwork can be selected on a small scale. From a methodological perspective, it is necessary to realize the organic combination of multidisciplinary methods and form a methodological unity rather than merely presenting scattered methods.

This book re-examines the process of urbanization in China based on simplifying and reconstructing the theory of production of space. According to these typical cases, from the macro-scale Yangtze River Delta urban agglomeration and Jiangsu Province, to the meso-scale New Urban District, State-level New Area and National High-Tech Industrial Development Zone, to the small-scale Xianlin University Town, Shanghai's Wujing Township and Heping Village, we represent the different but connected processes and landscapes of urbanization. These different scales are like different sides of a prism, reflecting the interaction between urbanization and production of space from multiple angles and aspects and then depicting the interlaced relationship between power, capital, class, and space in the process of urbanization.

At the macro-scale, the dominance of power, large-scale space, and capital expansion constitute the developmental mode of urbanization in the Yangtze River Delta and Jiangsu Province. However, this developmental mode does not narrow the gap between the urban and rural areas, and instead increased regional and social imbalances. At the meso-scale, the National High-Tech Development Zone has a significant spatial and temporal imbalance. As a by-product of this unique form of spatial organization, the unconventional speed and scale of urbanization coupled with unique policy privileges have formed a boom in the construction of development zones. Different levels of government dominate the production of space in these new districts and special zones. Geographical space has become a field of power games and rights struggles. This is an interweaving process of power reconstruction and spatial reorganization, which ultimately leads to new imbalances or inadequate development.

At the micro-scale, university towns reflect the absence of university culture and the multi-scalar separations of time, space and society. At a more microscopic scale of townships and villages, power drives the separation of social spaces. Bottom-up community self-organization is suppressed by top-down policy forces, while capital-fueled rising house prices widen the gap between locals and outsiders. As shown by a typical case of village in Shanghai, the high-tension lines of power and capital, like the physical high-tension line, make villagers lose their sense of belonging. The large-scale spatial expansion driven by capital and power has brought about great spatial imbalances and has also weakened the social space that people depend on, ultimately leading to lost (urban and rural) culture.

From global, national, regional, local to place and even body, the production of space is reflected at multiple scales. There are different research focuses at different scales, and different scales are intertwined to form a "symphony" of production of space (Table 11.1). The focus of the "global-state" scale is to recognize the overall structure. Therefore, the research should be combined with remote sensing images and GIS to carry out comprehensive research based on multi-country comparison, macro-policy together with institutional analysis, and statistical analysis. The "region-city" scale focuses on spatiotemporal evolution. Key concepts such as cycles of capital, power and class changes, and scale conversions need to be explored in a targeted manner. Through quantitative research such as the construction and evaluation of the index system, quantitative analysis, spatial analysis, and policy analysis, we can clarify the process of the transfer and function of capital within and between

Table 11.1 Key points of methodology for multi-scale spatial production research

Scale	Focus	Key concept	Method
Global and state	Overall structure	Financial capitalism; uneven development and space–time compression; neoliberal critique	Comprehensive research: RS, GIS, multi-country comparison, macro policy and institutional analysis, statistical analysis, etc.
Region and city	Spatiotemporal evolution	Three cycles of capital; power and class change; scale transformation	Quantitative research: indicator system, econometric analysis, spatial analysis, policy analysis, etc.
Community and individual	Everyday life	Power and rights; body politics; mirror-image; mediocrity and miracle	Qualitative research: participation in observation, in-depth interview, social investigation, historical analysis, combination of literature and art, etc.

Scale transformation: "scale up" and "scale down"
Scale resilience: "imaging the big from the small" and "convert complex to simple"
Scale correlation: the process of interaction between time, space and society (class, power, capital)

cities, as well as the shaping of social space by power. The core of the "community-individual" scale is to critique everyday life and explore the relationship between social space and individual life. We need to use qualitative research methods such as observation, in-depth interviews, and social surveys to explore issues such as the game of interests between different social classes, the relationship between everyday life and social spaces, and the right to the city.

11.4 The Future of Urbanization: From People-Oriented to Humanism

With the rapid urbanization process after the reform and opening up, large-scale land use, high energy and resource consumption, and the immense urban–rural gap have made China's sustainable development face severe challenges. Entering into the new era, the factors affecting sustainable development and urbanization have undergone great changes. Especially since 2020, the drastic changes in globalization, intelligence, and super-epidemic and social mobility have had a significant impact on the spatiotemporal pattern and development trend of urbanization. The year 2020 also marks the final year of China's first round of new-type urbanization planning. In this context, it is necessary to rethink the future path to urbanization.

The new type of urbanization emphasizes "people-oriented," and its core goal is to realize the urbanization of people. However, to some extent, the status quo of urbanization characterized and driven by production of space is in conflict with, or even contradicts, the "people-oriented" new type of urbanization. People's rights to urban life are still dominated by factors such as power and capital. The "people" who should be the main body of society is in a passive, aphasic, and marginalized situation as long as urbanization dominated by the logic of production of space. The private and everyday life spaces of individuals is squeezed by power and capital, and the issue of free movement and the right to city is still unclear. It is also difficult for individuals to participate in the planning and decision-making process of urban and rural construction. These factors have seriously restricted the healthy development of urbanization.

Faced with the goal of sustainable development and building new urban civilization, the basic orientation of future urbanization should be humanism that emphasizes the harmony between society, technology and nature, rather than a narrow sense of people-oriented concept.

Humanism has two levels: at the human (individual) level, there is a close relationship between individual values, personality, feelings, psychology, and space and place; at the cultural level, nature, history, philosophy, social, and organizational behavior are also closely related to geographic space. From a socio-spatial perspective, the integration of approaches from different disciplines (such as economics, management, sociology, and geography) is expected to advance research on social transformation issues. Through class analysis from sociology and multi-scale perspectives or methods from geography, the research focuses on the space of all social classes, especially the space of vulnerable groups. The combination of social justice and spatial justice not only highlights the value concern of humanities and social sciences but also the method orientation of academic or scientific research. Bridging to link different socio-spatial scales (such as urban–rural, region, and community) together by balancing the gaps between social classes will be the focus of theory and practice.

Our ultimate goal, as Marx pointed out, is not only to interpret the world, but also to change the world. Historically, perhaps change may be driven by external forces, but radical change ultimately comes from internal itself. The practice or reconstruction of the theory of production of space is a process from the outside to the inside and then to the unification of inside and outside. This requires more collective and public efforts and participation. It probably comes down to self-practice and self-actualization. Humanistic urbanization values both social and technological dimensions, as well as the natural and ecological environment and sustainable goals. Its direction is individual freedom and the highly developed human civilization formed by urbanization. This is not an ideal or impossible goal but an urgent task in reality that we must face and act to seek changes. The interaction and transformation between theory and practice are never-ending processes. Urbanization and production of space are not the discourse of "grand narratives" that have nothing to do with individuals. On the contrary, they are the structural force that affects everyone and the reality that dominate our lives. Therefore, if we have seen the fundamental problems and forces in history and reality, then the paths and solutions to them are clear. That is all in our own hands and acting. Everything including ourselves is becoming, and reality is changing and can be changed. If we have been braved enough to recognize this, then there is only one thing to do, and that is to take the same courage to practice or change it.